3rd Grade Multiplication Games

by: Laura Putman, Bright Minds Engaged

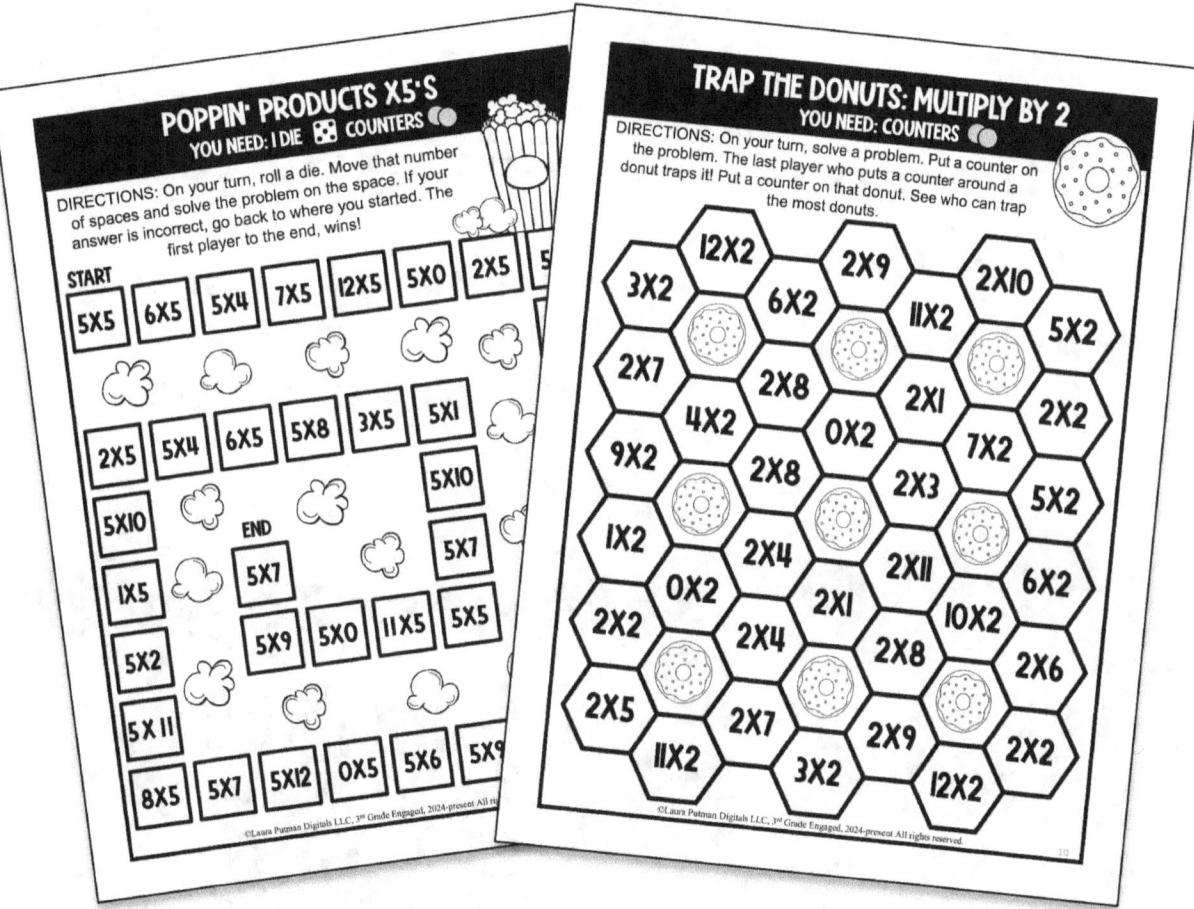

© Laura Putman Digitals, LLC, 3rd Grade Engaged, 2024-present, All rights reserved.

All images created by Laura Putman Digitals LLC. All rights reserved. No part of this publication may be reproduced, distributed, or transmitted in any form or by any means. This includes photocopying, recording, or other electronic or mechanical methods without prior permission of the publisher, except in the case of brief quotations embodied in critical reviews and other noncommercial uses permitted by copyright law.

Clipart by Educlips – www.educlips.com
Partyhead Graphics
Johnny's Clips
Hidesy's Clipart

No part of this product maybe used or reproduced for commercial use.

Contact the author :
laura@thirdgradeengaged.com

MULTIPLICATION FACTS

Parent Strategy Guide

Get the most out of this workbook!

✓ Tips for Success

✓ Strategy Posters

✓ Progress Chart

Scan here to get it!

The guide to helping your child master multiplication facts

By Laura Putman, M.Ed

© Laura Putman, Bright Minds Engaged, 2026-present, All rights reserved.

TABLE OF CONTENTS

p. 4-8 – multiplying by 0 and 1

p. 9-17 – multiplying by 2

p. 18-26 – multiplying by 3

p. 27-35 – multiplying by 4

p. 36-44 – multiplying by 5

p. 45-53 – multiplying by 6

p. 54-62 – multiplying by 7

p. 63-71 – multiplying by 8

p. 72-80 – multiplying by 9

p. 81-87 – multiplying by 10

p. 88-95 – multiplying by 11

p. 96-103 – multiplying by 12

p. 104-114 – multiplying mixed facts

p. 115-122 – multiplying by multiples of 10

p. 123-140 – Properties of Multiplication

p. 141-151 – repeated addition and arrays

ROLL & SOLVE: MULTIPLY BY 0 & 1

YOU NEED: 1 DIE CRAYONS

DIRECTIONS: Assign one player even numbers on the die and the other player odd numbers. Take turns rolling. If a player rolls one of their numbers on the die, they solve the next problem under that die and color the space. If they do not roll one of their numbers, their turn is over. See who can fill their columns first!

⚀	⚁	⚂	⚃	⚄	⚅
1X1	0X5	7X1	0X10	8X0	2X1
0X12	1X11	10X1	1X8	1X3	0X0
5X1	1X8	0X4	1X3	12X1	7X1
0X2	0X1	9X1	4X0	0X7	0X9
4X1	3X0	5X1	11X0	0X3	1X6
1X0	12X1	6X0	1X3	1X6	11X0

TRAP THE SPIDERS: MULTIPLY BY 0 & 1
YOU NEED: COUNTERS

DIRECTIONS: On your turn, solve a problem. Put a counter on the problem. The last player who puts a counter around a spider traps it! Put a counter on that spider. See who can trap the most spiders.

	12X1		1X9		1X10	
3X0		6X0		11X0		5X1
1X7		0X8		1X1		2X0
	4X1		0X0		7X0	
9X0		1X8		1X3		5X0
1X0		0X4		1X11		6X1
	0X0		0X1		10X0	
2X1		1X4		1X8		0X6
1X5		1X7		0X9		2X1
	11X0		3X1		12X1	

©Laura Putman Digitals LLC, 3rd Grade Engaged, 2024-present All rights reserved.

TIC-TAC-TOE: MULTIPLY BY 0 & 1

DIRECTIONS: Play a game of tic-tac-toe! Before you mark a space as yours, you must solve the problem in that space.

0X0	12X1	1X1
0X7	0X9	11X1
1X10	0X8	1X5

6X0	7X1	1X2
0X1	8X0	3X1
9X1	1X0	0X4

1X9	5X0	1X10
0X1	1X8	2X0
1X3	0X7	0X12

4X1	1X3	12X0
5X1	1X12	0X11
7X1	1X0	8X0

0X7	0X1	0X5
1X6	10X0	1X4
2X1	0X11	3X1

1X6	0X3	4X1
11X1	10X1	0X0
0X2	6X1	9X1

FOUR PROBLEMS IN A ROW: MULTIPLY BY 0 & 1
YOU NEED: CRAYONS OR COUNTERS

DIRECTIONS: On your turn, solve a problem and color it or cover it. Each player uses a different color. The first player to get 4 in a row wins!

0 X 2	0 X 1	9 X 0	6 X 1	0 X 1	1 X 7
5 X 1	10 X 0	12 X 0	4 X 1	0 X 8	1 X 2
7 X 0	1 X 8	0 X 0	1 X 3	1 X 4	1 X 0
6 X 1	0 X 11	1 X 7	6 X 0	1 X 1	11 X 1
0 X 2	5 X 1	9 X 0	1 X 8	10 X 0	3 X 1
1 X 12	11 X 0	6 X 1	9 X 0	1 X 5	4 X 0

MATH BATTLE: X0 & 1
YOU NEED: 1 DIE 🎲 COUNTERS ●●

DIRECTIONS: On your turn, roll a die. Move that number of spaces and solve the problem on the space. If your answer is incorrect, go back to where you started. The first player to the end, wins!

START

0X3		0X0	8X1	0X7		0X9	1X6

END

7X1		9X1		10X1		3X1	
0X11		0X5		0X11		1X10	
4X1		1X2		2X1		4X0	
0X6		0X8		1X0		1X1	
10X1		12X1		1X0		7X0	
0X12		0X9		1X6		1X11	
1X1	5X0	4X1		8X1	3X0	1X2	

©Laura Putman Digitals LLC, 3rd Grade Engaged, 2024-present All rights reserved.

ROLL & SOLVE: MULTIPLY BY 2
YOU NEED: 1 DIE 🎲 CRAYONS 🖍️

DIRECTIONS: Assign one player even numbers on the die and the other player odd numbers. Take turns rolling. If a player rolls one of their numbers on the die, they solve the next problem under that die and color the space. If they do not roll one of their numbers, their turn is over. See who can fill their columns first!

⚀	⚁	⚂	⚃	⚄	⚅
2X1	2X5	7X2	2X0	8X2	2X3
2X12	2X11	10X2	2X8	2X9	0X2
5X2	2X8	2X4	3X2	12X2	7X2
1X2	1X2	2X0	4X2	2X7	2X9
4X2	9X2	5X2	2X11	2X2	2X6
2X10	12X2	6X2	2X3	2X6	11X2

©Laura Putman Digitals LLC, 3rd Grade Engaged, 2024-present All rights reserved.

TRAP THE DONUTS: MULTIPLY BY 2
YOU NEED: COUNTERS

DIRECTIONS: On your turn, solve a problem. Put a counter on the problem. The last player who puts a counter around a donut traps it! Put a counter on that donut. See who can trap the most donuts.

	12X2		2X9		2X10	
3X2		6X2		11X2		5X2
	🍩		🍩		🍩	
2X7		2X8		2X1		2X2
	4X2		0X2		7X2	
9X2		2X8		2X3		5X2
	🍩		🍩		🍩	
1X2		2X4		2X11		6X2
	0X2		2X1		10X2	
2X2		2X4		2X8		2X6
	🍩		🍩		🍩	
2X5		2X7		2X9		2X2
	11X2		3X2		12X2	

TIC-TAC-TOE: MULTIPLY BY 2

DIRECTIONS: Play a game of tic-tac-toe! Before you mark a space as yours, you must solve the problem in that space.

0X2	12X2	2X1
2X7	2X9	11X2
2X10	2X8	2X5

6X2	7X2	2X2
0X2	8X2	3X2
9X2	1X2	2X4

2X9	5X2	2X10
2X1	2X8	2X2
2X3	2X7	2X12

4X2	2X3	12X2
5X2	2X12	2X11
7X2	1X2	8X2

2X7	0X2	2X5
2X6	10X2	2X4
2X2	2X11	3X2

2X6	2X3	4X2
11X2	10X2	2X0
2X2	6X2	9X2

SPIN A PROBLEM: MULTIPLY BY 2
YOU NEED: PAPERCLIP 📎 PENCIL ✏️ CRAYONS

DIRECTIONS: Use a paperclip and pencil to make a spinner. On your turn, spin the paperclip. Multiply that number by 2 and color the product in the table. Each player uses a different color. If the product is not open, your turn is over. See who can solve the most problems!

2	14	24	8	4	16
8	18	12	10	20	6
22	4	10	16	2	8
12	24	10	20	18	22
24	6	18	4	14	12

PIG IN A PEN: MULTIPLY BY 2
YOU NEED: PAPERCLIP ✏ PENCIL ✏

DIRECTIONS: On their turn, each player spins a number. Multiply the number by 2 and say the answer. If the answer is correct, draw a line to connect 2 dots. When a player completes a box, they write their initial in the box. At the end of the game, boxes are worth 1 point, and boxes with a pig in them are worth 5 points!

ARRAYS? HOORAY! MULTIPLY BY 2
YOU NEED: CRAYONS   2 DICE

DIRECTIONS: On their turn, each player rolls both dice. Add the dice together and multiply by 2. Draw an array for the problem. Each player uses a different color to make their arrays. Write the multiplication problem inside the array. If your array won't fit, your turn is over. When no more arrays can be made, the game is over. Whoever makes the most arrays, wins!

MULTIPLES OF 2 HUNT
YOU NEED: CRAYONS

Help the skiers make it down the mountain. Count by 2's to follow a path. If you get to 24, start at 2 again. Color or dab the spaces until you get to the bottom.

START →

		2	15	12	81	21	38	56	44
		12	4	20	40	24	63	75	42
		22	16	6	8	14	5	80	79
		6	65	16	40	10	36	52	9
24	15	80	66	28	44	45	10	15	12	8	78
64	10	12	28	88	24	50	35	18	16	14	13
47	37	24	4	2	41	22	20	24	60	18	40
20	8	6	20	1	5	60	63	65	27	7	38
66	10	23	40	15	20	19	72
41	21	12	14	18	63	22	18
28	14	81	16	22	50	25	24
78	35	42	40	71	8	55	2

→ END

FOUR PROBLEMS IN A ROW: MULTIPLY BY 2

YOU NEED: CRAYONS OR COUNTERS

DIRECTIONS: On your turn, solve a problem and color it or cover it. Each player uses a different color. The first player to get 4 in a row wins!

2 X 2	0 X 2	9 X 2	6 X 2	2 X 1	2 X 7
5 X 2	10 X 2	12 X 2	4 X 2	2 X 8	2 X 2
7 X 2	2 X 8	2 X 0	2 X 3	2 X 4	1 X 2
6 X 2	2 X 11	2 X 7	6 X 2	1 X 2	11 X 2
2 X 2	5 X 2	9 X 2	2 X 8	10 X 2	3 X 2
2 X 12	11 X 2	6 X 2	9 X 2	2 X 5	4 X 2

MULTIPLICATION MYSTERY X2'S

YOU NEED: 1 DIE COUNTERS

DIRECTIONS: On your turn, roll a die. Move that number of spaces and solve the problem on the space. If your answer is incorrect, go back to where you started. The first player to the end, wins!

START

2X3		2X0	8X2	2X7		2X9	2X6
7X2		9X2		10X2		3X2	**END**
2X11		2X5		2X11		2X10	
4X2		2X2		2X2		4X2	
2X6		2X8		2X0		2X1	
10X2		12X2		1X2		7X2	
2X12		2X9		2X6		2X11	
1X2	5X2	4X2		8X2	3X2	2X2	

©Laura Putman Digitals LLC, 3rd Grade Engaged, 2024-present All rights reserved.

ROLL & SOLVE: MULTIPLY BY 3

YOU NEED: 1 DIE CRAYONS

DIRECTIONS: Assign one player even numbers on the die and the other player odd numbers. Take turns rolling. If a player rolls one of their numbers on the die, they solve the next problem under that die and color the space. If they do not roll one of their numbers, their turn is over. See who can fill their columns first!

⚀	⚁	⚂	⚃	⚄	⚅
3X1	3X5	7X3	3X10	8X3	2X3
3X12	3X11	10X3	3X8	3X3	0X3
5X3	3X8	3X4	3X3	12X3	7X3
3X2	1X3	9X3	4X3	3X7	3X9
4X3	3X3	5X3	11X3	2X3	3X6
3X0	12X3	6X3	3X3	3X6	11X3

TRAP THE YETIS: MULTIPLY BY 3
YOU NEED: COUNTERS

DIRECTIONS: On your turn, solve a problem. Put a counter on the problem. The last player who puts a counter around a yeti traps it! Put a counter on that yeti. See who can trap the most yetis.

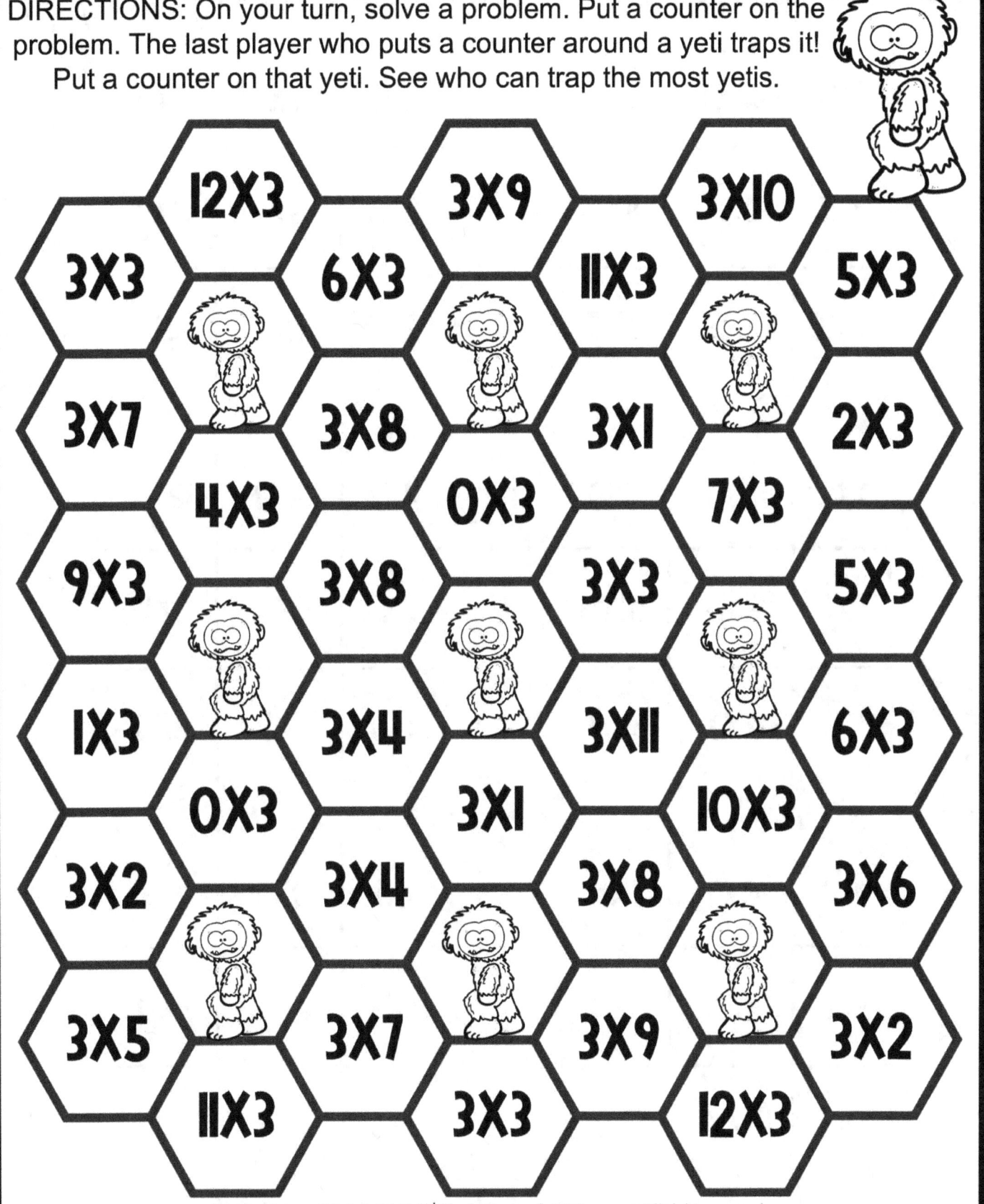

TIC-TAC-TOE: MULTIPLY BY 3

DIRECTIONS: Play a game of tic-tac-toe! Before you mark a space as yours, you must solve the problem in that space.

0X3	12X3	3X1
3X7	3X9	11X3
3X10	3X8	3X5

6X3	7X3	2X3
0X3	8X3	3X3
9X3	1X3	3X4

3X9	5X3	3X10
3X1	3X8	3X2
3X3	3X7	3X12

4X3	3X3	12X3
5X3	3X12	3X11
7X3	1X3	8X3

3X7	0X3	3X5
3X6	10X3	3X4
3X2	3X11	3X3

3X6	3X3	4X3
11X3	10X3	3X0
2X3	6X3	9X3

SPIN A PROBLEM: MULTIPLY BY 3
YOU NEED: PAPERCLIP 📎 PENCIL ✏️ CRAYONS

DIRECTIONS: Use a paperclip and pencil to make a spinner. On your turn, spin the paperclip. Multiply that number by 3 and color the product in the table. Each player uses a different color. If the product is not open, your turn is over. See who can solve the most problems!

3	21	36	12	6	24
12	27	18	15	30	9
33	6	3	24	3	12
18	36	15	30	27	33
30	9	27	6	21	18

PIG IN A PEN: MULTIPLY BY 3
YOU NEED: PAPERCLIP 🖇 PENCIL ✏

DIRECTIONS: On their turn, each player spins a number. Multiply the number by 3 and say the answer. If the answer is correct, draw a line to connect 2 dots. When a player completes a box, they write their initial in the box. At the end of the game, boxes are worth 1 point, and boxes with a pig in them are worth 5 points!

ARRAYS? HOORAY! MULTIPLY BY 3
YOU NEED: CRAYONS   2 DICE

DIRECTIONS: On their turn, each player rolls both dice. Add the dice together and multiply by 3. Draw an array for the problem. Each player uses a different color to make their arrays. Write the multiplication problem inside the array. If your array won't fit, your turn is over. When no more arrays can be made, the game is over. Whoever makes the most arrays, wins!

MULTIPLES OF 3 HUNT
YOU NEED: CRAYONS

Help the bookworm get to the books. Count by 3's to follow a path. If you get to 36, start at 3 again. Color or dab the spaces until you get to the bottom.

		START →	3	6	13	81	21	38	56	43
			12	10	9	40	24	63	75	42
			22	16	12	25	30	5	80	79
			6	65	56	15	35	36	52	9
24	15	80	66	23	21	18	10	42	71	8	78
14	10	74	16	24	80	50	35	54	49	62	13
47	37	33	27	84	41	16	55	56	60	18	40
20	36	28	30	83	5	60	63	65	27	7	38
66	3	25	40	15	24	27	72				
6	21	30	18	56	21	74	30				
28	9	81	35	18	13	33	75				
78	21	12	15	71	78	55	36	→ END			

FOUR PROBLEMS IN A ROW: MULTIPLY BY 3

YOU NEED: CRAYONS OR COUNTERS

DIRECTIONS: On your turn, solve a problem and color it or cover it. Each player uses a different color. The first player to get 4 in a row wins!

2 X 3	0 X 3	9 X 3	6 X 3	3 X 1	3 X 7
5 X 3	10 X 3	12 X 3	4 X 3	3 X 8	3 X 2
7 X 3	3 X 8	3 X 0	3 X 3	3 X 4	1 X 3
6 X 3	3 X 11	3 X 7	6 X 3	1 X 3	11 X 3
3 X 2	5 X 3	9 X 3	3 X 8	10 X 3	3 X 3
3 X 12	11 X 3	6 X 3	9 X 3	3 X 5	4 X 3

FOOTBALL FACTS X3'S

YOU NEED: 1 DIE  COUNTERS

DIRECTIONS: On your turn, roll a die. Move that number of spaces and solve the problem on the space. If your answer is incorrect, go back to where you started. The first player to the end, wins!

START

| 5X3 | 6X3 | 3X4 | 7X3 | 12X3 | 3X0 | 2X3 | 3X9 |

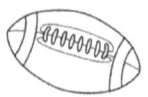   

8X3

| 2X3 | 3X4 | 6X3 | 3X8 | 3X3 | 3X1 | | 3X3 |

 22

| 3X10 | | | | 3X10 | | 10X3 |

44 **END**

| 1X3 | 3X7 | | 4X3 | | 1X3 |

| 3X2 | | 3X9 | 3X0 | 11X3 | 5X3 | | 11X3 |

| 3X11 | | | | | | 5X3 |

| 8X3 | 3X7 | 3X12 | 0X3 | 3X6 | 3X9 | 3X3 | 4X3 |

©Laura Putman Digitals LLC, 3rd Grade Engaged, 2024-present All rights reserved.

ROLL & SOLVE: MULTIPLY BY 4
YOU NEED: 1 DIE · CRAYONS

DIRECTIONS: Assign one player even numbers on the die and the other player odd numbers. Take turns rolling. If a player rolls one of their numbers on the die, they solve the next problem under that die and color the space. If they do not roll one of their numbers, their turn is over. See who can fill their columns first!

⚀	⚁	⚂	⚃	⚄	⚅
4X1	4X5	7X4	2X4	8X4	4X3
4X12	4X11	10X4	4X8	4X9	0X4
5X4	4X8	4X4	3X4	12X4	7X4
1X4	1X4	4X0	4X4	4X7	4X9
4X4	9X4	5X4	4X11	4X2	4X6
4X10	12X4	6X4	4X3	4X6	11X4

TRAP THE UFOS: MULTIPLY BY 4
YOU NEED: COUNTERS

DIRECTIONS: On your turn, solve a problem. Put a counter on the problem. The last player who puts a counter around a UFO traps it! Put a counter on that UFO. See who can trap the most UFOs.

12X4	4X9	4X10
4X3	6X4	11X4	5X4
4X7	4X8	4X1	2X4
4X4	0X4	7X4
9X4	4X8	3X4	5X4
1X4	4X4	4X11	6X4
0X4	4X1	10X4
4X2	3X4	4X8	4X6
4X5	4X7	4X9	4X2
11X4	4X3	12X4

TIC-TAC-TOE: MULTIPLY BY 4

DIRECTIONS: Play a game of tic-tac-toe! Before you mark a space as yours, you must solve the problem in that space.

0X4	12X4	4X1
4X7	4X9	11X4
4X10	4X8	4X5

6X4	7X4	2X4
0X4	8X4	4X3
9X4	1X4	4X4

4X9	5X4	4X10
4X1	4X8	4X2
3X4	4X7	4X12

4X4	4X3	12X4
5X4	4X12	4X11
7X4	1X4	8X4

4X7	0X4	4X5
4X6	10X4	4X4
4X2	4X11	3X4

4X6	4X3	4X4
11X4	10X4	4X0
2X4	6X4	9X4

SPIN A PROBLEM: MULTIPLY BY 4
YOU NEED: PAPERCLIP / PENCIL / CRAYONS

DIRECTIONS: Use a paperclip and pencil to make a spinner. On your turn, spin the paperclip. Multiply that number by 4 and color the product in the table. Each player uses a different color. If the product is not open, your turn is over. See who can solve the most problems!

4	28	48	16	8	32
16	36	24	20	40	12
44	8	16	32	4	16
24	48	20	40	36	44
32	12	36	8	28	24

PIG IN A PEN: MULTIPLY BY 4
YOU NEED: PAPERCLIP 🖇 PENCIL ✏

DIRECTIONS: On their turn, each player spins a number. Multiply the number by 4 and say the answer. If the answer is correct, draw a line to connect 2 dots. When a player completes a box, they write their initial in the box. At the end of the game, boxes are worth 1 point, and boxes with a pig in them are worth 5 points!

ARRAYS? HOORAY! MULTIPLY BY 4
YOU NEED: CRAYONS 2 DICE

DIRECTIONS: On their turn, each player rolls both dice. Add the dice together and multiply by 4. Draw an array for the problem. Each player uses a different color to make their arrays. Write the multiplication problem inside the array. If your array won't fit, your turn is over. When no more arrays can be made, the game is over. Whoever makes the most arrays, wins!

MULTIPLES OF 4 HUNT
YOU NEED: CRAYONS

Help the kids make it to the ice cream. Count by 4's to follow a path. If you get to 48, start at 4 again. Color or dab the spaces until you get to the bottom.

			START →	4	15	12	81	21	38	56	75
				8	10	20	40	24	18	75	42
				22	12	23	25	32	5	80	79
				16	65	56	40	35	36	52	9
24	15	80	66	27	20	44	10	32	71	40	78
64	14	74	16	28	80	24	28	54	49	44	19
47	37	24	96	32	24	30	55	12	60	4	48
20	48	28	36	2	5	20	16	65	8	7	38
66	23	25	4	40	10	19	72				
41	21	30	18	56	44	48	6				
28	64	81	35	45	40	4	75				
78	35	42	40	71	78	55	8	END →			

FOUR PROBLEMS IN A ROW: MULTIPLY BY 4

YOU NEED: CRAYONS OR COUNTERS

DIRECTIONS: On your turn, solve a problem and color it or cover it. Each player uses a different color. The first player to get 4 in a row wins!

2 X 4	0 X 4	9 X 4	6 X 4	4 X 1	4 X 7
5 X 4	10 X 4	12 X 4	4 X 4	4 X 8	4 X 2
7 X 4	4 X 8	4 X 0	3 X 4	4 X 4	1 X 4
6 X 4	4 X 11	4 X 7	6 X 4	1 X 4	11 X 4
4 X 2	5 X 4	9 X 4	4 X 8	10 X 4	4 X 3
4 X 12	11 X 4	6 X 4	9 X 4	4 X 5	4 X 4

©Laura Putman Digitals LLC, 3rd Grade Engaged, 2024-present All rights reserved.

MULTIPLICATION MYSTERY X4'S

YOU NEED: 1 DIE COUNTERS

DIRECTIONS: On your turn, roll a die. Move that number of spaces and solve the problem on the space. If your answer is incorrect, go back to where you started. The first player to the end, wins!

START

4X3		4X0	8X4	4X7		4X9	4X6
7X4		9X4		10X4		3X4	**END**
4X11		4X5		4X11		4X10	
4X4		2X4		2X4		4X4	
4X6		4X8		4X0		4X1	
10X4		12X4		1X4		7X4	
4X12		4X9		4X6		4X11	
1X4	5X4	4X4		8X4	3X4	4X2	

©Laura Putman Digitals LLC, 3rd Grade Engaged, 2024-present All rights reserved.

ROLL & SOLVE: MULTIPLY BY 5

YOU NEED: 1 DIE CRAYONS

DIRECTIONS: Assign one player even numbers on the die and the other player odd numbers. Take turns rolling. If a player rolls one of their numbers on the die, they solve the next problem under that die and color the space. If they do not roll one of their numbers, their turn is over. See who can fill their columns first!

⚀	⚁	⚂	⚃	⚄	⚅
5X1	5X5	7X5	5X10	8X5	2X5
5X12	5X11	10X5	5X8	3X5	0X5
5X5	5X8	5X4	5X3	12X5	7X5
5X2	1X5	9X5	4X5	5X7	5X9
4X5	5X2	5X5	11X5	2X5	5X6
5X0	12X5	6X5	5X3	5X6	11X5

TRAP THE SHARKS: MULTIPLY BY 5
YOU NEED: COUNTERS

DIRECTIONS: On your turn, solve a problem. Put a counter on the problem. The last player who puts a counter around a shark traps it! Put a counter on that shark. See who can trap the most sharks.

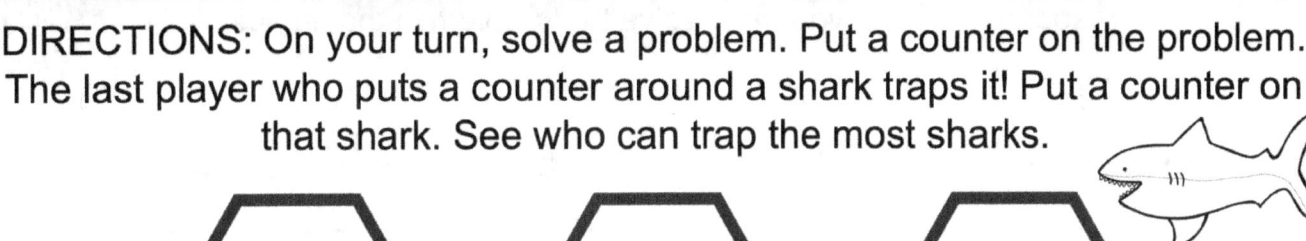

	12X5		5X9		5X10	
5X3		6X5		11X5		5X5
5X7		5X8		5X1		2X5
	5X4		0X5		7X5	
9X5		5X8		3X5		5X5
1X5		4X5		5X12		6X5
	0X5		5X1		10X5	
5X2		3X5		5X8		5X6
5X5		5X7		5X9		5X2
	11X5		5X3		12X5	

TIC-TAC-TOE: MULTIPLY BY 5

DIRECTIONS: Play a game of tic-tac-toe! Before you mark a space as yours, you must solve the problem in that space.

0X5	12X5	5X1
5X7	5X9	11X5
5X10	5X8	5X5

6X5	7X5	2X5
0X5	8X5	5X3
9X5	1X5	5X4

5X9	5X5	5X10
5X1	5X8	5X2
3X5	5X7	5X12

4X5	5X3	12X5
5X5	5X12	5X11
7X5	1X5	8X5

5X7	0X5	5X5
5X6	10X5	4X5
5X2	5X11	3X5

5X6	5X3	5X4
11X5	10X5	5X0
2X5	6X5	9X5

SPIN A PROBLEM: MULTIPLY BY 5
YOU NEED: PAPERCLIP / PENCIL / CRAYONS

DIRECTIONS: Use a paperclip and pencil to make a spinner. On your turn, spin the paperclip. Multiply that number by 5 and color the product in the table. Each player uses a different color. If the product is not open, your turn is over. See who can solve the most problems!

5	35	60	20	10	40
20	45	30	25	50	15
55	10	15	40	5	20
30	60	25	50	45	55
25	15	45	10	35	30

PIG IN A PEN: MULTIPLY BY 5
YOU NEED: PAPERCLIP ✏ PENCIL ✏

DIRECTIONS: On their turn, each player spins a number. Multiply the number by 5 and say the answer. If the answer is correct, draw a line to connect 2 dots. When a player completes a box, they write their initial in the box. At the end of the game, boxes are worth 1 point, and boxes with a pig in them are worth 5 points!

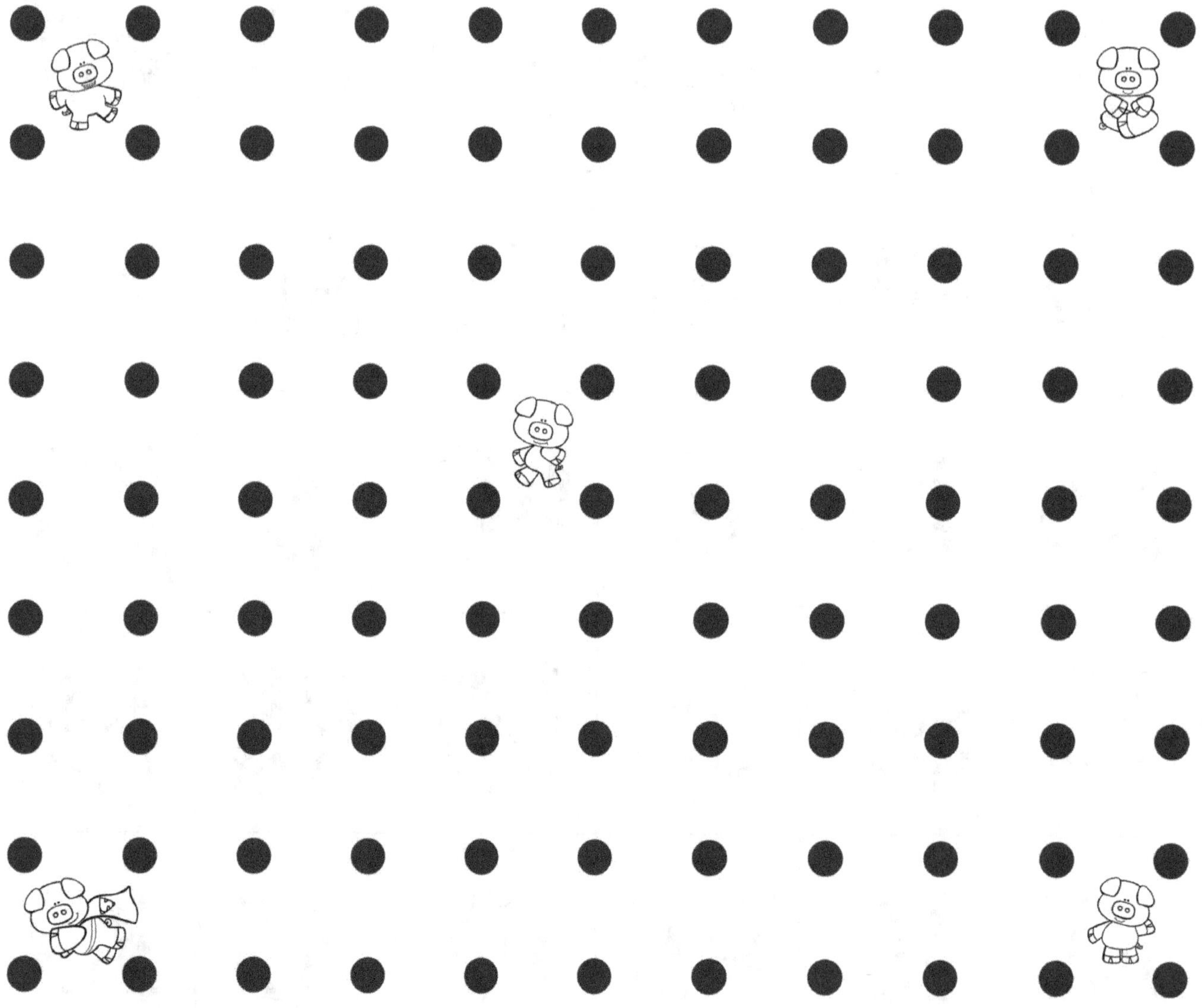

ARRAYS? HOORAY! MULTIPLY BY 5
YOU NEED: CRAYONS   2 DICE

DIRECTIONS: On their turn, each player rolls both dice. Add the dice together and multiply by 5. Draw an array for the problem. Each player uses a different color to make their arrays. Write the multiplication problem inside the array. If your array won't fit, your turn is over. When no more arrays can be made, the game is over. Whoever makes the most arrays, wins!

MULTIPLES OF 5 HUNT
YOU NEED: CRAYONS

Help the panda find its friends. Count by 5's to follow a path. If you get to 60, start at 5 again. Color or dab the spaces until you get to the bottom.

			START →	5	15	12	81	21	38	56	25
				12	10	20	40	24	63	75	42
				22	16	23	25	30	5	80	79
				6	65	56	40	35	36	52	9
24	15	80	66	27	44	45	10	42	71	8	78
64	10	74	16	88	80	50	35	54	49	62	13
47	37	24	96	84	41	16	55	56	60	18	40
20	48	28	20	83	5	60	63	65	27	7	38
66	23	25	40	15	10	19	72				
41	21	30	18	56	63	74	18				
28	52	81	35	45	50	72	75				
78	35	42	40	71	78	55	60	→ END			

FOUR PROBLEMS IN A ROW: MULTIPLY BY 5
YOU NEED: CRAYONS

DIRECTIONS: On your turn, solve a problem and color it or cover it. Each player uses a different color. The first player to get 4 in a row wins!

2 X 5	0 X 5	9 X 5	6 X 5	5 X 1	5 X 7
5 X 5	10 X 5	12 X 5	5 X 4	5 X 8	5 X 2
7 X 5	5 X 8	5 X 0	3 X 5	4 X 5	1 X 5
6 X 5	5 X 11	5 X 7	6 X 5	1 X 5	11 X 5
5 X 2	5 X 5	9 X 5	5 X 8	10 X 5	5 X 3
5 X 12	11 X 5	6 X 5	9 X 5	5 X 5	4 X 5

POPPIN' PRODUCTS X5'S

YOU NEED: 1 DIE 🎲 COUNTERS ⚫⚫

DIRECTIONS: On your turn, roll a die. Move that number of spaces and solve the problem on the space. If your answer is incorrect, go back to where you started. The first player to the end, wins!

START

| 5X5 | 6X5 | 5X4 | 7X5 | 12X5 | 5X0 | 2X5 | 5X9 |

| | | | | | | | 8X5 |

| 2X5 | 5X4 | 6X5 | 5X8 | 3X5 | 5X1 | | 5X3 |

| 5X10 | | | | | 5X10 | | 10X5 |

END

| 1X5 | | 5X7 | | | 5X7 | | 1X5 |

| 5X2 | | 5X9 | 5X0 | 11X5 | 5X5 | | 11X5 |

| 5X11 | | | | | | | 5X5 |

| 8X5 | 5X7 | 5X12 | 0X5 | 5X6 | 5X9 | 5X3 | 4X5 |

©Laura Putman Digitals LLC, 3rd Grade Engaged, 2024-present All rights reserved.

ROLL & SOLVE: MULTIPLY BY 6
YOU NEED: 1 DIE 🎲 CRAYONS

DIRECTIONS: Assign one player even numbers on the die and the other player odd numbers. Take turns rolling. If a player rolls one of their numbers on the die, they solve the next problem under that die and color the space. If they do not roll one of their numbers, their turn is over. See who can fill their columns first!

⚀	⚁	⚂	⚃	⚄	⚅
6X1	6X5	7X6	2X6	8X6	6X3
6X12	6X11	10X6	6X8	6X9	0X6
5X6	6X8	4X6	3X6	12X6	7X6
1X6	1X6	6X0	6X4	6X7	6X9
6X4	9X6	5X6	6X11	6X2	6X6
6X10	12X6	6X6	6X3	6X6	11X6

TRAP THE CREATURES: MULTIPLY BY 6
YOU NEED: COUNTERS

DIRECTIONS: On your turn, solve a problem. Put a counter on the problem. The last player who puts a counter around a creature traps it! Put a counter on that creature. See who can trap the most creatures.

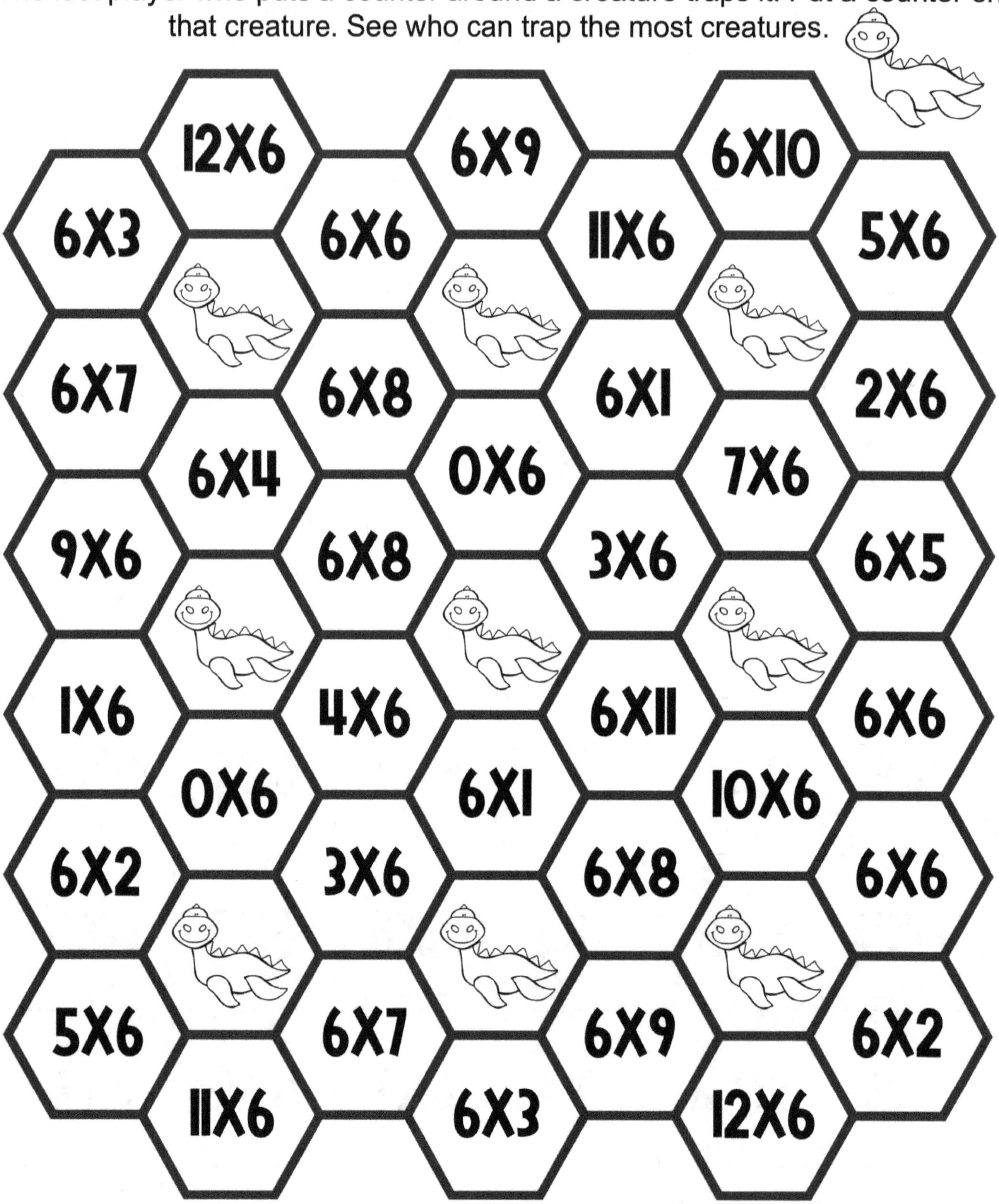

	12X6		6X9		6X10	
6X3		6X6		11X6		5X6
6X7		6X8		6X11		2X6
	6X4		0X6		7X6	
9X6		6X8		3X6		6X5
1X6		4X6		6X11		6X6
	0X6		6X1		10X6	
6X2		3X6		6X8		6X6
5X6		6X7		6X9		6X2
	11X6		6X3		12X6	

©Laura Putman Digitals LLC, 3rd Grade Engaged, 2024-present All rights reserved.

TIC-TAC-TOE: MULTIPLY BY 6

DIRECTIONS: Play a game of tic-tac-toe! Before you mark a space as yours, you must solve the problem in that space.

0X6	12X6	6X1
6X7	6X9	11X6
6X10	6X8	6X5

6X6	7X6	2X6
0X6	8X6	6X3
9X6	1X6	6X4

6X9	5X6	6X10
6X1	6X8	6X2
3X6	6X7	6X12

4X6	6X3	12X6
5X6	6X12	6X11
7X6	1X6	8X6

6X7	0X6	6X5
6X6	10X6	4X6
6X2	6X11	3X6

6X6	6X3	6X4
11X6	10X6	6X0
2X6	6X6	9X6

SPIN A PROBLEM: MULTIPLY BY 6
YOU NEED: PAPERCLIP / PENCIL / CRAYONS

DIRECTIONS: Use a paperclip and pencil to make a spinner. On your turn, spin the paperclip. Multiply that number by 6 and color the product in the table. Each player uses a different color. If the product is not open, your turn is over. See who can solve the most problems!

6	42	72	24	12	48
24	54	36	30	60	18
66	12	18	48	6	24
36	72	30	60	54	66
30	18	54	12	42	36

PIG IN A PEN: MULTIPLY BY 6
YOU NEED: PAPERCLIP / PENCIL

DIRECTIONS: On their turn, each player spins a number. Multiply the number by 6 and say the answer. If the answer is correct, draw a line to connect 2 dots. When a player completes a box, they write their initial in the box. At the end of the game, boxes are worth 1 point, and boxes with a pig in them are worth 5 points!

ARRAYS? HOORAY! MULTIPLY BY 6
YOU NEED: CRAYONS   2 DICE

DIRECTIONS: On their turn, each player rolls both dice. Add the dice together and multiply by 6. Draw an array for the problem. Each player uses a different color to make their arrays. Write the multiplication problem inside the array. If your array won't fit, your turn is over. When no more arrays can be made, the game is over. Whoever makes the most arrays, wins!

MULTIPLES OF 6 HUNT
YOU NEED: CRAYONS

Help the raccoon win the snowball fight! Count by 6's to follow a path. If you get to 72, start at 6 again. Color or dab the spaces until you get to the bottom.

			START →	6	15	12	18	21	38	56	33
				12	14	20	40	24	63	12	42
				36	18	24	28	32	9	80	79
				6	65	22	30	35	36	52	9
24	15	80	6	27	44	33	36	42	71	8	78
64	10	74	16	8	18	72	35	54	48	62	13
47	37	24	60	40	41	5	60	54	6	88	40
21	48	28	7	32	72	66	63	11	27	42	38
18	23	18	40	6	32	19	72				
41	21	79	18	56	12	18	24				
28	52	81	49	15	70	30	75				
78	35	42	35	71	78	77	36	END →			

FOUR PROBLEMS IN A ROW: MULTIPLY BY 6
YOU NEED: CRAYONS OR COUNTERS

DIRECTIONS: On your turn, solve a problem and color it or cover it. Each player uses a different color. The first player to get 4 in a row wins!

2 X 6	0 X 6	9 X 6	6 X 6	6 X 1	6 X 7
6 X 5	10 X 6	12 X 6	6 X 4	6 X 8	6 X 2
7 X 6	6 X 8	6 X 0	3 X 6	4 X 6	1 X 6
6 X 6	6 X 11	6 X 7	6 X 6	1 X 6	11 X 6
6 X 2	5 X 6	9 X 6	6 X 8	10 X 6	6 X 3
6 X 12	11 X 6	6 X 6	9 X 6	6 X 5	4 X 6

©Laura Putman Digitals LLC, 3rd Grade Engaged, 2024-present All rights reserved.

MATH BATTLE: X6'S

YOU NEED: 1 DIE 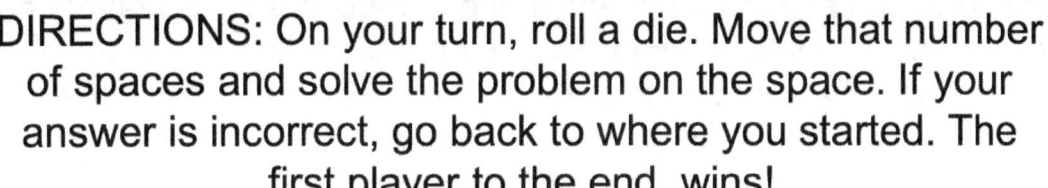 COUNTERS

DIRECTIONS: On your turn, roll a die. Move that number of spaces and solve the problem on the space. If your answer is incorrect, go back to where you started. The first player to the end, wins!

START

6X3		6X0	8X6	6X7		6X9	6X6
7X6		9X6		10X6		3X6	**END**
6X11		6X5		6X11		6X10	
4X6		6X2		2X6		4X6	
6X6		6X8		6X0		6X1	
10X6		12X6		1X6		7X6	
6X12		6X9		6X6		6X11	
1X6	5X6	4X6		8X6	3X6	6X2	

©Laura Putman Digitals LLC, 3rd Grade Engaged, 2024-present All rights reserved.

ROLL & SOLVE: MULTIPLY BY 7

YOU NEED: 1 DIE 🎲 CRAYONS

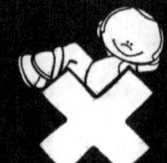

DIRECTIONS: Assign one player even numbers on the die and the other player odd numbers. Take turns rolling. If a player rolls one of their numbers on the die, they solve the next problem under that die and color the space. If they do not roll one of their numbers, their turn is over. See who can fill their columns first!

⚀	⚁	⚂	⚃	⚄	⚅
7X1	5X7	7X7	7X10	8X7	2X7
7X12	7X11	10X7	7X8	3X7	0X7
5X7	7X8	7X4	7X3	12X7	7X7
7X2	1X7	9X7	4X7	7X7	7X9
4X7	7X2	7X5	11X7	2X7	7X6
7X0	12X7	6X7	7X3	7X6	11X7

TRAP THE TREASURE: MULTIPLY BY 7
YOU NEED: COUNTERS

DIRECTIONS: On your turn, solve a problem. Put a counter on the problem. The last player who puts a counter around a chest traps it! Put a counter on that chest. See who can trap the most chests.

- 12X7, 7X9, 7X10
- 7X3, 6X7, 11X7, 5X7
- 7X7, 7X8, 7X1, 2X7
- 7X4, 0X7, 7X7
- 9X7, 7X8, 3X7, 7X5
- 1X7, 4X7, 7X12, 6X7
- 0X7, 7X1, 10X7
- 7X2, 3X7, 7X8, 7X6
- 5X7, 7X7, 7X9, 7X2
- 11X7, 7X3, 12X7

TIC-TAC-TOE: MULTIPLY BY 7

DIRECTIONS: Play a game of tic-tac-toe! Before you mark a space as yours, you must solve the problem in that space.

0X7	12X7	7X1
7X7	7X9	11X7
7X10	7X8	7X5

7X6	7X7	2X7
0X7	8X7	7X3
9X7	1X7	7X4

7X9	5X7	7X10
7X1	7X8	7X2
3X7	7X7	7X12

4X7	7X3	12X7
5X7	7X12	7X11
7X7	1X7	8X7

7X7	0X7	7X5
6X7	10X7	4X7
7X2	7X11	3X7

6X7	7X3	7X4
11X7	10X7	7X0
2X7	7X6	9X7

SPIN A PROBLEM: MULTIPLY BY 7
YOU NEED: PAPERCLIP / PENCIL / CRAYONS

DIRECTIONS: Use a paperclip and pencil to make a spinner. On your turn, spin the paperclip. Multiply that number by 7 and color the product in the table. Each player uses a different color. If the product is not open, your turn is over. See who can solve the most problems!

7	49	84	28	14	56
28	63	42	35	70	21
77	14	21	56	7	28
42	84	35	70	63	77
35	21	63	14	49	42

PIG IN A PEN: MULTIPLY BY 7
YOU NEED: PAPERCLIP 📎 PENCIL ✏️

DIRECTIONS: On their turn, each player spins a number. Multiply the number by 7 and say the answer. If the answer is correct, draw a line to connect 2 dots. When a player completes a box, they write their initial in the box. At the end of the game, boxes are worth 1 point, and boxes with a pig in them are worth 5 points!

ARRAYS? HOORAY! MULTIPLY BY 7
YOU NEED: CRAYONS   2 DICE

DIRECTIONS: On their turn, each player rolls both dice. Add the dice together and multiply by 7. Draw an array for the problem. Each player uses a different color to make their arrays. Write the multiplication problem inside the array. If your array won't fit, your turn is over. When no more arrays can be made, the game is over. Whoever makes the most arrays, wins!

MULTIPLES OF 7 HUNT
YOU NEED: CRAYONS

Help the kids win the water fight! Count by 7's to follow a path. If you get to 84, start at 7 again. Color or dab the spaces until you get to the bottom.

			7	15	12	81	21	38	56	48
			16	14	21	40	24	63	75	42
			36	24	23	28	32	9	80	79
			6	65	56	40	35	36	52	9
24	15	80	66	27	44	64	13	42	71	8	78
64	10	74	16	88	80	72	35	54	49	62	13
47	37	24	96	84	41	16	52	56	66	88	40
21	48	28	7	83	77	70	63	11	27	78	38
66	23	14	40	48	32	19	72				
41	21	79	18	56	63	74	18				
28	52	81	49	15	70	72	75				
78	35	42	35	71	78	77	84				

FOUR PROBLEMS IN A ROW: MULTIPLY BY 7
YOU NEED: CRAYONS OR COUNTERS

DIRECTIONS: On your turn, solve a problem and color it or cover it. Each player uses a different color. The first player to get 4 in a row wins!

2 X 7	0 X 7	9 X 7	6 X 7	7 X 1	7 X 7
7 X 5	10 X 7	12 X 7	7 X 4	7 X 8	7 X 2
7 X 7	7 X 8	7 X 0	3 X 7	4 X 7	1 X 7
6 X 7	7 X 11	7 X 7	7 X 6	1 X 7	11 X 7
7 X 2	5 X 7	9 X 7	7 X 8	10 X 7	7 X 3
7 X 12	11 X 7	7 X 6	9 X 7	7 X 5	4 X 7

FOOTBALL FACTS X7'S

YOU NEED: 1 DIE COUNTERS

DIRECTIONS: On your turn, roll a die. Move that number of spaces and solve the problem on the space. If your answer is incorrect, go back to where you started. The first player to the end, wins!

START

| 5X7 | 6X7 | 7X4 | 7X7 | 12X7 | 7X0 | 2X7 | 7X9 |

8X7

| 2X7 | 7X4 | 6X7 | 7X8 | 3X7 | 7X1 | | 7X3 |

 22

| 7X10 | | | | 7X10 | | 10X7 |

44   **END**

| 1X7 | | 7X7 | | | 4X7 | | 1X7 |

| 7X2 | | 7X9 | 7X0 | 11X7 | 5X7 | | 11X7 |

| 7X11 | | | | | | | 5X7 |

| 8X7 | 7X7 | 7X12 | 0X7 | 7X6 | 7X9 | 7X3 | 4X7 |

ROLL & SOLVE: MULTIPLY BY 8
YOU NEED: 1 DIE CRAYONS

DIRECTIONS: Assign one player even numbers on the die and the other player odd numbers. Take turns rolling. If a player rolls one of their numbers on the die, they solve the next problem under that die and color the space. If they do not roll one of their numbers, their turn is over. See who can fill their columns first!

⚀	⚁	⚂	⚃	⚄	⚅
8X1	8X5	7X8	2X8	8X8	8X3
8X12	8X11	10X8	8X8	8X9	0X8
5X8	8X8	4X8	3X8	12X8	7X8
1X8	1X8	8X0	8X4	8X7	8X9
8X4	9X8	5X8	8X11	8X2	8X6
8X10	12X8	8X6	8X3	6X8	11X8

TRAP THE MONKEYS: MULTIPLY BY 8
YOU NEED: COUNTERS

DIRECTIONS: On your turn, solve a problem. Put a counter on the problem. The last player who puts a counter around a monkey traps it! Put a counter on that monkey. See who can trap the most monkeys.

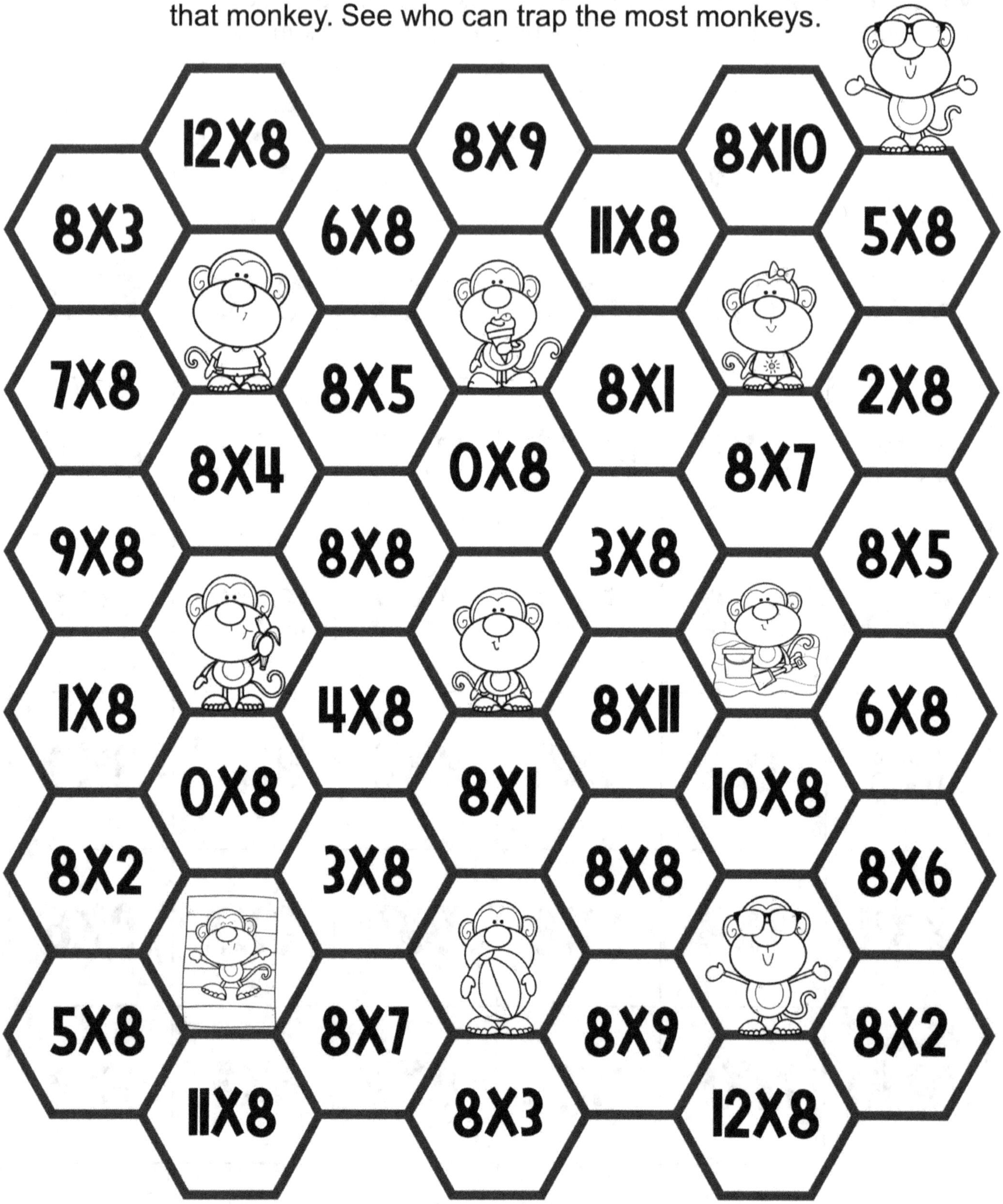

	12X8		8X9		8X10	
8X3		6X8		11X8		5X8
7X8		8X5		8X1		2X8
	8X4		0X8		8X7	
9X8		8X8		3X8		8X5
1X8		4X8		8X11		6X8
	0X8		8X1		10X8	
8X2		3X8		8X8		8X6
5X8		8X7		8X9		8X2
	11X8		8X3		12X8	

©Laura Putman Digitals LLC, 3rd Grade Engaged, 2024-present All rights reserved.

TIC-TAC-TOE: MULTIPLY BY 8

DIRECTIONS: Play a game of tic-tac-toe! Before you mark a space as yours, you must solve the problem in that space.

0X8	12X8	8X1
8X7	8X9	11X8
8X10	8X8	8X5

8X6	7X8	2X8
0X8	8X8	8X3
9X8	1X8	8X4

8X9	5X8	8X10
8X1	8X8	8X2
3X8	8X7	8X12

4X8	8X3	12X8
5X8	8X12	8X11
7X8	1X8	8X8

8X7	0X8	8X5
6X8	10X8	4X8
8X2	8X11	3X8

6X8	8X3	8X4
11X8	10X8	8X0
2X8	8X6	8X7

SPIN A PROBLEM: MULTIPLY BY 8
YOU NEED: PAPERCLIP 📎 PENCIL ✏️ CRAYONS

DIRECTIONS: Use a paperclip and pencil to make a spinner. On your turn, spin the paperclip. Multiply that number by 8 and color the product in the table. Each player uses a different color. If the product is not open, your turn is over. See who can solve the most problems!

8	56	96	32	16	64
32	72	48	40	80	24
88	16	24	64	8	64
48	96	40	80	72	88
40	24	72	16	56	48

PIG IN A PEN: MULTIPLY BY 8
YOU NEED: PAPERCLIP 📎 PENCIL ✏️

DIRECTIONS: On their turn, each player spins a number. Multiply the number by 8 and say the answer. If the answer is correct, draw a line to connect 2 dots. When a player completes a box, they write their initial in the box. At the end of the game, boxes are worth 1 point, and boxes with a pig in them are worth 5 points!

ARRAYS? HOORAY! MULTIPLY BY 8
YOU NEED: CRAYONS   2 DICE

DIRECTIONS: On their turn, each player rolls both dice. Add the dice together and multiply by 8. Draw an array for the problem. Each player uses a different color to make their arrays. Write the multiplication problem inside the array. If your array won't fit, your turn is over. When no more arrays can be made, the game is over. Whoever makes the most arrays, wins!

MULTIPLES OF 8 HUNT
YOU NEED: CRAYONS

Help the kid get to the candy! Count by 8's to follow a path. If you get to 96, start at 8 again. Color or dab the spaces until you get to the bottom.

				8	16	12	81	21	38	56	55
				18	14	24	40	24	63	75	42
				36	24	23	32	38	9	80	79
				6	65	56	42	40	36	52	9
24	48	80	66	27	44	64	13	42	48	8	78
63	56	40	44	88	80	72	35	54	49	56	13
64	72	24	32	84	16	19	96	56	66	88	64
21	88	80	7	24	77	8	63	88	80	72	38
96	94	14	40	48	36	19	72				
8	24	79	18	63	77	74	18				
16	52	32	56	64	80	89	94				
78	40	48	35	71	78	88	96				

FOUR PROBLEMS IN A ROW: MULTIPLY BY 8
YOU NEED: CRAYONS OR COUNTERS

DIRECTIONS: On your turn, solve a problem and color it or cover it. Each player uses a different color. The first player to get 4 in a row wins!

2 X 8	0 X 8	9 X 8	6 X 8	8 X 1	8 X 7
8 X 5	10 X 8	12 X 8	8 X 4	8 X 8	8 X 2
7 X 8	8 X 8	8 X 0	3 X 8	4 X 8	1 X 8
6 X 8	8 X 11	8 X 7	8 X 6	1 X 8	11 X 8
8 X 2	5 X 8	9 X 8	8 X 8	10 X 8	8 X 3
8 X 12	11 X 8	8 X 6	9 X 8	8 X 5	4 X 8

MULTIPLICATION MYSTERY X8'S

YOU NEED: 1 DIE COUNTERS

DIRECTIONS: On your turn, roll a die. Move that number of spaces and solve the problem on the space. If your answer is incorrect, go back to where you started. The first player to the end, wins!

START

8X3		8X0	8X8	7X8		8X9	8X6
7X8		9X8		10X8		3X8	END
8X11		8X5		8X11		8X10	
4X8		2X8		2X8		4X8	
8X6		8X8		8X0		8X1	
10X8		12X8		1X8		7X8	
8X12		8X9		8X6		8X11	
1X8	5X8	8X4		8X8	3X8	8X2	

ROLL & SOLVE: MULTIPLY BY 9

YOU NEED: 1 DIE CRAYONS

DIRECTIONS: Assign one player even numbers on the die and the other player odd numbers. Take turns rolling. If a player rolls one of their numbers on the die, they solve the next problem under that die and color the space. If they do not roll one of their numbers, their turn is over. See who can fill their columns first!

⚀	⚁	⚂	⚃	⚄	⚅
9X1	5X9	9X7	9X10	8X9	2X9
9X12	9X11	10X9	9X8	3X9	0X9
5X9	9X8	9X4	9X3	12X9	9X7
9X2	1X9	9X9	4X9	7X9	9X9
4X9	9X2	9X5	11X9	2X9	9X6
9X0	12X9	6X9	9X3	9X6	11X9

TRAP THE SUBMARINES: MULTIPLY BY 9
YOU NEED: COUNTERS

DIRECTIONS: On your turn, solve a problem. Put a counter on the problem. The last player who puts a counter around a sub traps it! Put a counter on that sub. See who can trap the most subs.

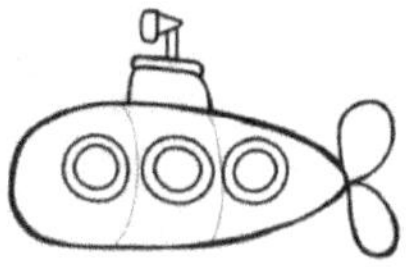

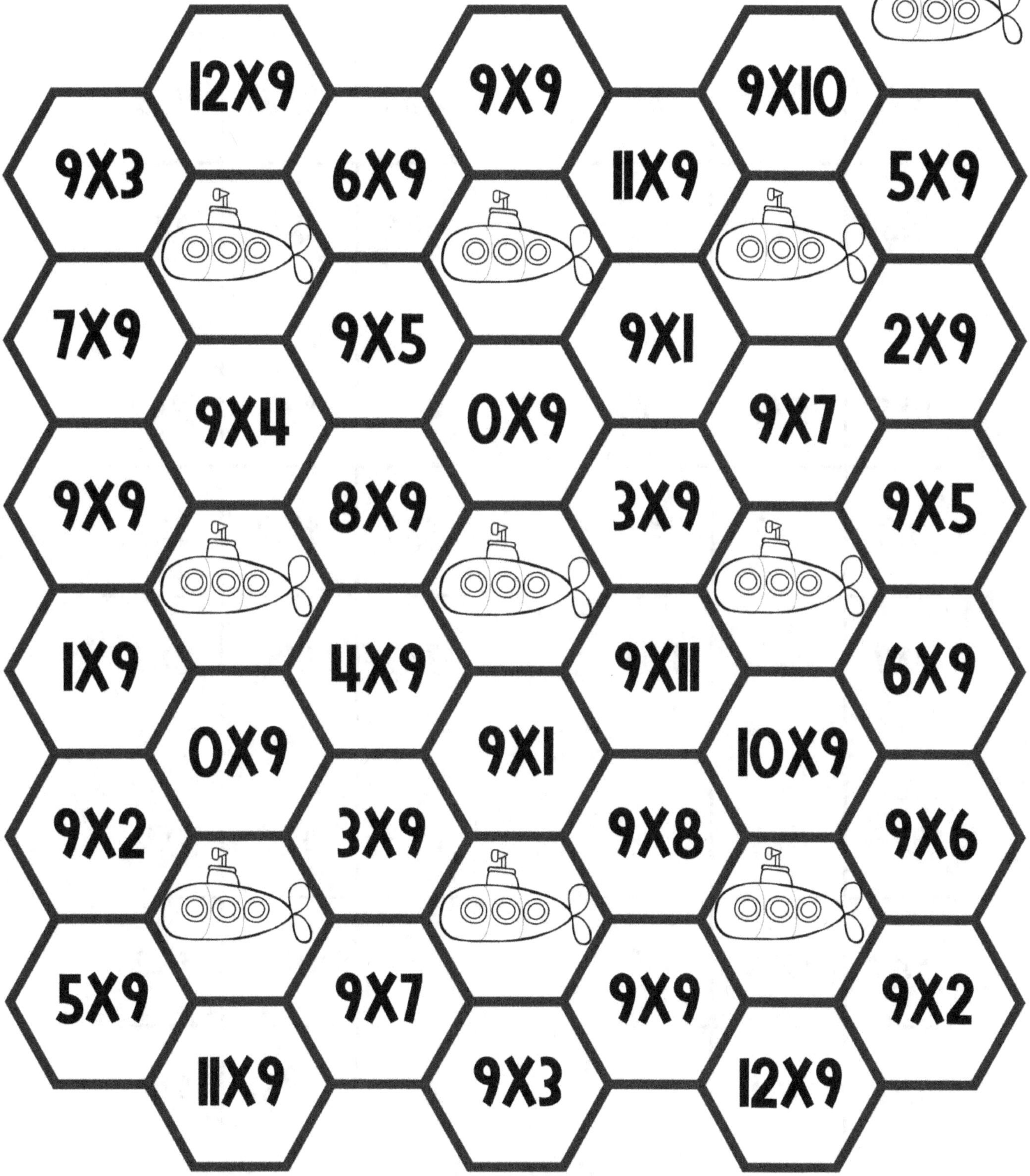

12X9	9X9	9X10
9X3	6X9	11X9	5X9
7X9	9X5	9X1	2X9
9X4	0X9	9X7
9X9	8X9	3X9	9X5
1X9	4X9	9X11	6X9
0X9	9X1	10X9
9X2	3X9	9X8	9X6
5X9	9X7	9X9	9X2
11X9	9X3	12X9

TIC-TAC-TOE: MULTIPLY BY 9

DIRECTIONS: Play a game of tic-tac-toe! Before you mark a space as yours, you must solve the problem in that space.

0X9	12X9	9X1
9X7	9X9	11X9
9X10	8X9	9X5

9X6	7X9	2X9
0X9	9X8	9X3
9X9	1X9	9X4

9X9	5X9	9X10
9X1	9X8	9X2
3X9	9X7	9X12

4X9	9X3	12X9
5X9	9X12	9X11
7X9	1X9	8X9

9X7	0X9	9X5
6X9	10X9	4X9
9X2	9X11	3X9

6X9	9X3	9X4
11X9	10X9	9X0
2X9	9X6	9X7

SPIN A PROBLEM: MULTIPLY BY 9
YOU NEED: PAPERCLIP / PENCIL / CRAYONS

DIRECTIONS: Use a paperclip and pencil to make a spinner. On your turn, spin the paperclip. Multiply that number by 9 and color the product in the table. Each player uses a different color. If the product is not open, your turn is over. See who can solve the most problems!

9	63	108	36	18	72
36	81	54	45	90	27
99	18	27	72	9	36
54	108	45	90	81	99
45	27	81	18	63	54

PIG IN A PEN: MULTIPLY BY 9
YOU NEED: PAPERCLIP 📎 PENCIL ✏️

DIRECTIONS: On their turn, each player spins a number. Multiply the number by 9 and say the answer. If the answer is correct, draw a line to connect 2 dots. When a player completes a box, they write their initial in the box. At the end of the game, boxes are worth 1 point, and boxes with a pig in them are worth 5 points!

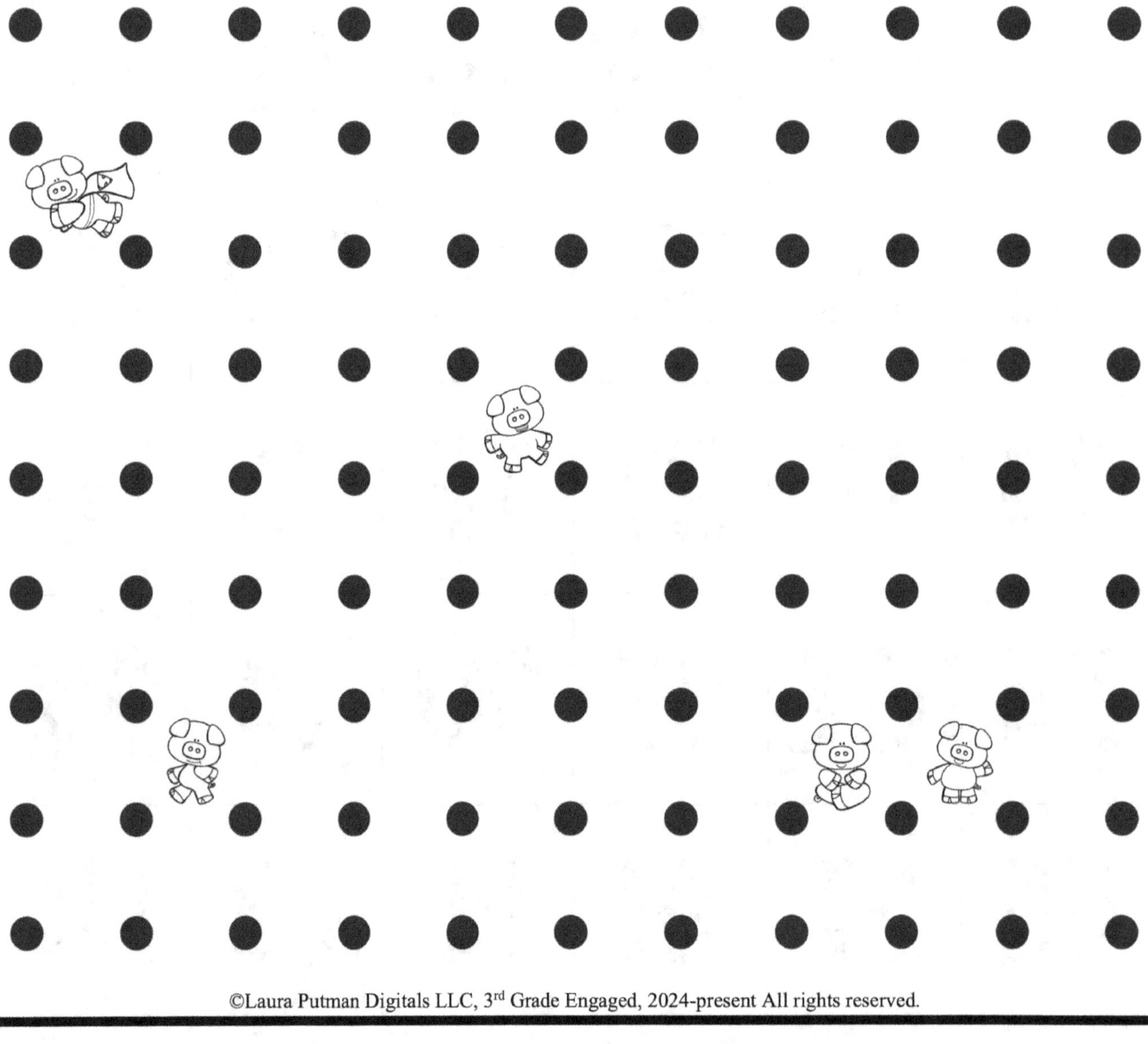

ARRAYS? HOORAY! MULTIPLY BY 9
YOU NEED: CRAYONS   2 DICE

DIRECTIONS: On their turn, each player rolls both dice. Add the dice together and multiply by 9. Draw an array for the problem. Each player uses a different color to make their arrays. Write the multiplication problem inside the array. If your array won't fit, your turn is over. When no more arrays can be made, the game is over. Whoever makes the most arrays, wins!

MULTIPLES OF 9 HUNT
YOU NEED: CRAYONS

Help make a touchdown! Count by 9's to follow a path. If you get to 108, start at 9 again. Color or dab the spaces until you get to the bottom.

			START →	9	16	12	81	21	38	56	19
				18	14	24	40	84	81	92	99
				36	27	23	32	72	9	90	108
				6	45	56	63	74	36	52	9
24	48	80	66	27	44	54	64	42	48	18	19
63	56	40	44	109	80	72	35	54	49	29	27
64	72	24	9	108	99	19	96	56	45	36	29
21	23	18	7	24	77	90	63	54	80	72	38
96	27	34	40	48	36	81	72				
8	36	45	76	72	81	74	18				
16	52	54	63	64	90	99	94				
78	40	48	35	71	78	110	108	→ END			

  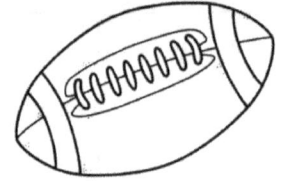 

FOUR PROBLEMS IN A ROW: MULTIPLY BY 9
YOU NEED: CRAYONS OR COUNTERS

DIRECTIONS: On your turn, solve a problem and color it or cover it. Each player uses a different color. The first player to get 4 in a row wins!

9 X 12	5 X 9	9 X 9	3 X 9	9 X 10	4 X 9
9 X 5	10 X 9	12 X 9	9 X 4	8 X 9	9 X 2
7 X 9	9 X 8	9 X 0	3 X 9	4 X 9	1 X 9
6 X 9	9 X 11	9 X 7	9 X 6	1 X 9	11 X 9
9 X 2	5 X 9	9 X 9	8 X 9	10 X 9	9 X 3
9 X 12	11 X 9	9 X 6	9 X 9	9 X 5	4 X 9

POPPIN' PRODUCTS X9'S

YOU NEED: 1 DIE  COUNTERS

DIRECTIONS: On your turn, roll a die. Move that number of spaces and solve the problem on the space. If your answer is incorrect, go back to where you started. The first player to the end, wins!

START

| 9X5 | 6X9 | 9X4 | 7X9 | 12X9 | 9X0 | 2X9 | 9X9 |

| | | | | | | | 8X9 |

| 2X9 | 9X4 | 6X9 | 9X8 | 3X9 | 9X1 | | 9X3 |

| 9X10 | | **END** | | 9X10 | | 10X9 |

| 1X9 | | 9X7 | | 9X7 | | 1X9 |

| 9X2 | | 9X9 | 9X0 | 11X9 | 5X9 | | 11X9 |

| 9X11 | | | | | | 9X5 |

| 8X9 | 9X7 | 9X12 | 0X9 | 9X6 | 9X9 | 9X3 | 4X9 |

©Laura Putman Digitals LLC, 3rd Grade Engaged, 2024-present All rights reserved.

TRAP THE ICE CREAM: MULTIPLY BY 10
YOU NEED: COUNTERS

DIRECTIONS: On your turn, solve a problem. Put a counter on the problem. The last player who puts a counter around a cone traps it! Put a counter on that cone. See who can trap the most cones.

12X10	10X9	10X10
10X3	6X10	11X10	5X10
7X10	10X5	10X1	2X10
	10X4	0X10	10X7	
9X10	10X8	3X10	10X5
1X10	4X10	10X11	6X10
	0X10	10X1	10X10	
10X2	3X10	10X8	10X6
5X10	10X7	10X9	10X2
	11X10	10X3	12X10	

SPIN A PROBLEM: MULTIPLY BY 10
YOU NEED: PAPERCLIP / PENCIL / CRAYONS

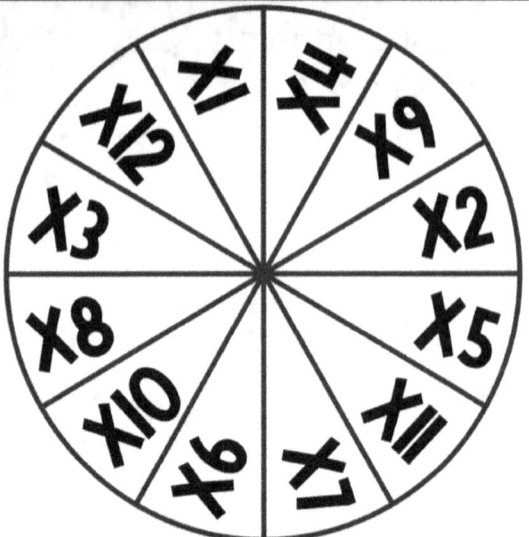

DIRECTIONS: Use a paperclip and pencil to make a spinner. On your turn, spin the paperclip. Multiply that number by 10 and color the product in the table. Each player uses a different color. If the product is not open, your turn is over. See who can solve the most problems!

10	70	120	40	20	80
40	90	60	50	100	30
110	20	30	80	10	40
60	120	50	100	90	110
50	30	90	20	70	60

PIG IN A PEN: MULTIPLY BY 10
YOU NEED: PAPERCLIP 📎 PENCIL ✏️

Spinner values: x1, x4, x9, x2, x5, x11, x7, x6, x10, x8, x3, x12

DIRECTIONS: On their turn, each player spins a number. Multiply the number by 10 and say the answer. If the answer is correct, draw a line to connect 2 dots. When a player completes a box, they write their initial in the box. At the end of the game, boxes are worth 1 point, and boxes with a pig in them are worth 5 points!

ARRAYS? HOORAY! MULTIPLY BY 10
YOU NEED: CRAYONS   2 DICE

DIRECTIONS: On their turn, each player rolls both dice. Add the dice together and multiply by 10. Draw an array for the problem. Each player uses a different color to make their arrays. Write the multiplication problem inside the array. If your array won't fit, your turn is over. When no more arrays can be made, the game is over. Whoever makes the most arrays, wins!

MULTIPLES OF 10 HUNT
YOU NEED: CRAYONS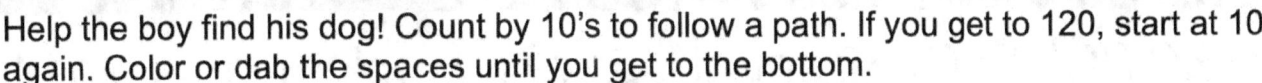

Help the boy find his dog! Count by 10's to follow a path. If you get to 120, start at 10 again. Color or dab the spaces until you get to the bottom.

START →	10	20	21	81	21	38	56	33
	18	14	30	20	84	81	92	99
	36	27	40	32	70	9	90	108
	60	50	56	72	74	110	52	9
24	48	80	80	70	40	54	60	40	62	11	19
63	56	90	44	109	80	72	35	54	49	29	27
64	110	100	9	108	99	10	60	112	45	36	29
120	23	18	7	50	24	90	30	54	80	72	38
10	20	30	40	60	36	81	72				
8	50	76	3	70	81	70	18				
16	102	54	80	64	90	3	94				
78	40	48	35	90	100	110	120	END →			

FOUR PROBLEMS IN A ROW: MULTIPLY BY 10

YOU NEED: CRAYONS OR COUNTERS

DIRECTIONS: On your turn, solve a problem and color it or cover it. Each player uses a different color. The first player to get 4 in a row wins!

10 X 12	5 X 10	10 X 9	3 X 10	10 X 10	4 X 10
10 X 5	10 X 10	12 X 10	10 X 4	8 X 10	10 X 2
7 X 10	10 X 8	10 X 0	3 X 10	4 X 10	1 X 10
6 X 10	10 X 11	10 X 7	10 X 6	1 X 10	11 X 10
10 X 2	5 X 10	9 X 10	8 X 10	10 X 10	10 X 3
10 X 12	11 X 10	10 X 6	10 X 9	10 X 5	4 X 10

©Laura Putman Digitals LLC, 3rd Grade Engaged, 2024-present All rights reserved.

MATH BATTLE: X10'S
YOU NEED: 1 DIE 🎲 COUNTERS ⬤⬤

DIRECTIONS: On your turn, roll a die. Move that number of spaces and solve the problem on the space. If your answer is incorrect, go back to where you started. The first player to the end, wins!

START

10X3		10X0	8X10	10X7		10X9	6X10
7X10		9X10		10X10		3X10	**END**
10X11		10X5		10X11		10X10	
4X10		10X2		2X10		4X10	
10X6		10X8		10X0		10X1	
10X10		12X10		1X10		7X10	
10X12		10X9		10X6		10X11	
1X10	5X10	4X10		8X10	3X10	10X2	

ROLL & SOLVE: MULTIPLY BY 10 & 11
YOU NEED: 1 DIE CRAYONS

DIRECTIONS: Assign one player even numbers on the die and the other player odd numbers. Take turns rolling. If a player rolls one of their numbers on the die, they solve the next problem under that die and color the space. If they do not roll one of their numbers, their turn is over. See who can fill their columns first!

⚀	⚁	⚂	⚃	⚄	⚅
8X11	10X5	7X10	2X10	10X8	11X3
10X12	11X11	10X11	8X11	11X9	0X10
5X10	10X8	4X10	3X10	12X11	7X11
1X11	1X10	11X0	4X11	10X7	11X4
10X4	9X10	5X11	10X10	11X2	10X6
8X10	12X11	10X6	11X3	6X11	10X9

TIC-TAC-TOE: MULTIPLY BY 10 & 11

DIRECTIONS: Play a game of tic-tac-toe! Before you mark a space as yours, you must solve the problem in that space.

0X10	12X11	9X11
10X7	11X9	11X9
9X10	8X11	10X5

11X6	7X10	2X11
0X11	10X8	11X3
10X9	1X11	10X4

11X9	5X10	11X10
10X1	11X8	10X2
3X10	11X7	10X12

4X11	10X3	12X11
5X11	10X12	11X11
7X10	1X11	8X11

10X7	0X10	11X5
6X10	10X11	4X10
11X2	10X11	11X9

6X10	10X3	11X4
11X11	10X10	10X0
2X11	10X6	11X7

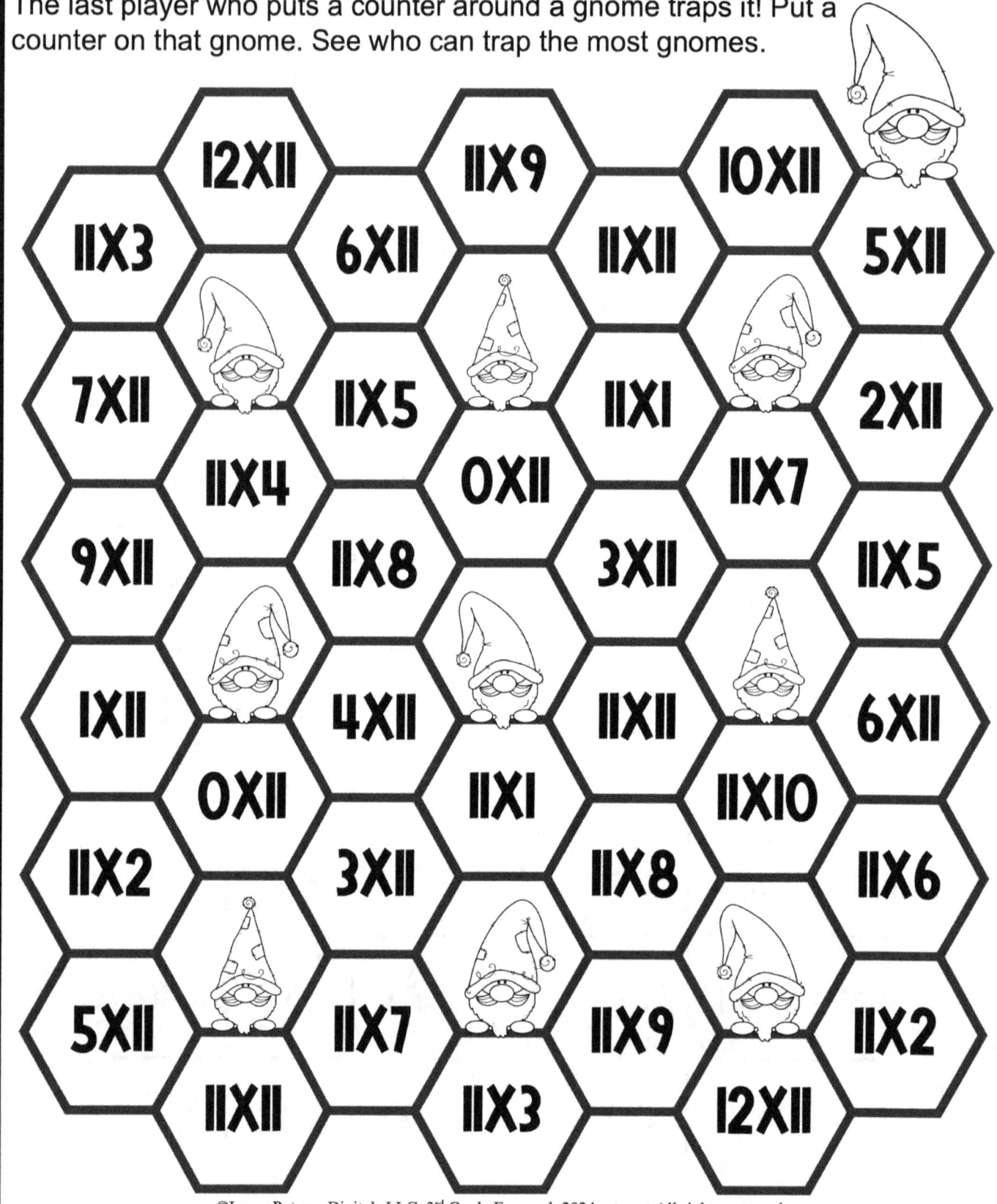

SPIN A PROBLEM: MULTIPLY BY 11
YOU NEED: PAPERCLIP / PENCIL / CRAYONS

DIRECTIONS: Use a paperclip and pencil to make a spinner. On your turn, spin the paperclip. Multiply that number by 11 and color the product in the table. Each player uses a different color. If the product is not open, your turn is over. See who can solve the most problems!

11	77	132	44	22	88
44	99	66	55	110	33
121	22	33	88	11	44
66	132	55	110	99	121
55	33	99	22	77	66

©Laura Putman Digitals LLC, 3rd Grade Engaged, 2024-present All rights reserved.

PIG IN A PEN: MULTIPLY BY 11
YOU NEED: PAPERCLIP 📎 PENCIL ✏️

DIRECTIONS: On their turn, each player spins a number. Multiply the number by 11 and say the answer. If the answer is correct, draw a line to connect 2 dots. When a player completes a box, they write their initial in the box. At the end of the game, boxes are worth 1 point, and boxes with a pig in them are worth 5 points!

MULTIPLES OF 11 HUNT
YOU NEED: CRAYONS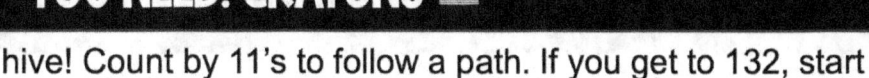

Help the honeybee get to its hive! Count by 11's to follow a path. If you get to 132, start at 11 again. Color or dab the spaces until you get to the bottom.

				11	16	12	81	21	38	56	60
				13	22	33	40	84	66	92	99
				36	27	23	44	55	77	90	108
				11	45	10	41	88	36	52	9
24	48	80	66	27	121	54	99	42	101	18	19
63	55	40	44	132	80	110	112	54	49	29	27
64	72	44	9	108	11	19	96	90	45	110	29
21	33	18	77	24	27	22	33	54	80	72	38
96	22	34	40	48	36	44	72				
8	30	44	76	72	55	74	77				
16	52	54	63	64	90	66	88				
88	40	48	35	71	78	110	99				

FOUR PROBLEMS IN A ROW: MULTIPLY BY 11
YOU NEED: CRAYONS OR COUNTERS

DIRECTIONS: On your turn, solve a problem and color it or cover it. Each player uses a different color. The first player to get 4 in a row wins!

11 X 12	5 X 11	11 X 9	3 X 11	10 X 11	4 X 11
2 X 11	0 X 11	11 X 9	6 X 11	11 X 1	11 X 7
7 X 11	11 X 8	11 X 0	3 X 11	4 X 11	1 X 11
6 X 11	11 X 11	11 X 7	11 X 6	1 X 11	11 X 11
11 X 2	5 X 11	9 X 11	8 X 11	11 X 10	11 X 3
11 X 12	11 X 11	11 X 6	11 X 9	11 X 5	4 X 11

MULTIPLICATION MYSTERY XII'S

YOU NEED: 1 DIE COUNTERS

DIRECTIONS: On your turn, roll a die. Move that number of spaces and solve the problem on the space. If your answer is incorrect, go back to where you started. The first player to the end, wins!

START

11X3		11X0	8X11	7X11		11X9	11X6 END
7X11		9X11		10X11		3X11	
11X11		11X5		11X11		11X10	
4X11		2X11		2X11		4X11	
11X6		11X8		11X0		11X1	
10X11		12X11		9X11		7X11	
11X12		11X9		11X6		11X11	
1X11	5X11	11X4		8X11	3X11	11X2	

©Laura Putman Digitals LLC, 3rd Grade Engaged, 2024-present All rights reserved.

ROLL & SOLVE: MULTIPLY BY 12
YOU NEED: 1 DIE 🎲 CRAYONS

DIRECTIONS: Assign one player even numbers on the die and the other player odd numbers. Take turns rolling. If a player rolls one of their numbers on the die, they solve the next problem under that die and color the space. If they do not roll one of their numbers, their turn is over. See who can fill their columns first!

⚀	⚁	⚂	⚃	⚄	⚅
12X1	5X12	12X7	12X10	8X12	2X12
12X12	12X11	10X12	12X8	3X12	0X12
5X12	12X8	12X4	12X3	12X12	12X7
12X2	1X12	12X9	4X12	7X12	12X9
4X12	12X2	12X5	11X12	2X12	12X6
12X0	12X12	6X12	12X3	12X6	11X12

TRAP THE MICE: MULTIPLY BY 12
YOU NEED: COUNTERS

DIRECTIONS: On your turn, solve a problem. Put a counter on the problem. The last player who puts a counter around a mouse traps it! Put a counter on that mouse. See who can trap the most mice.

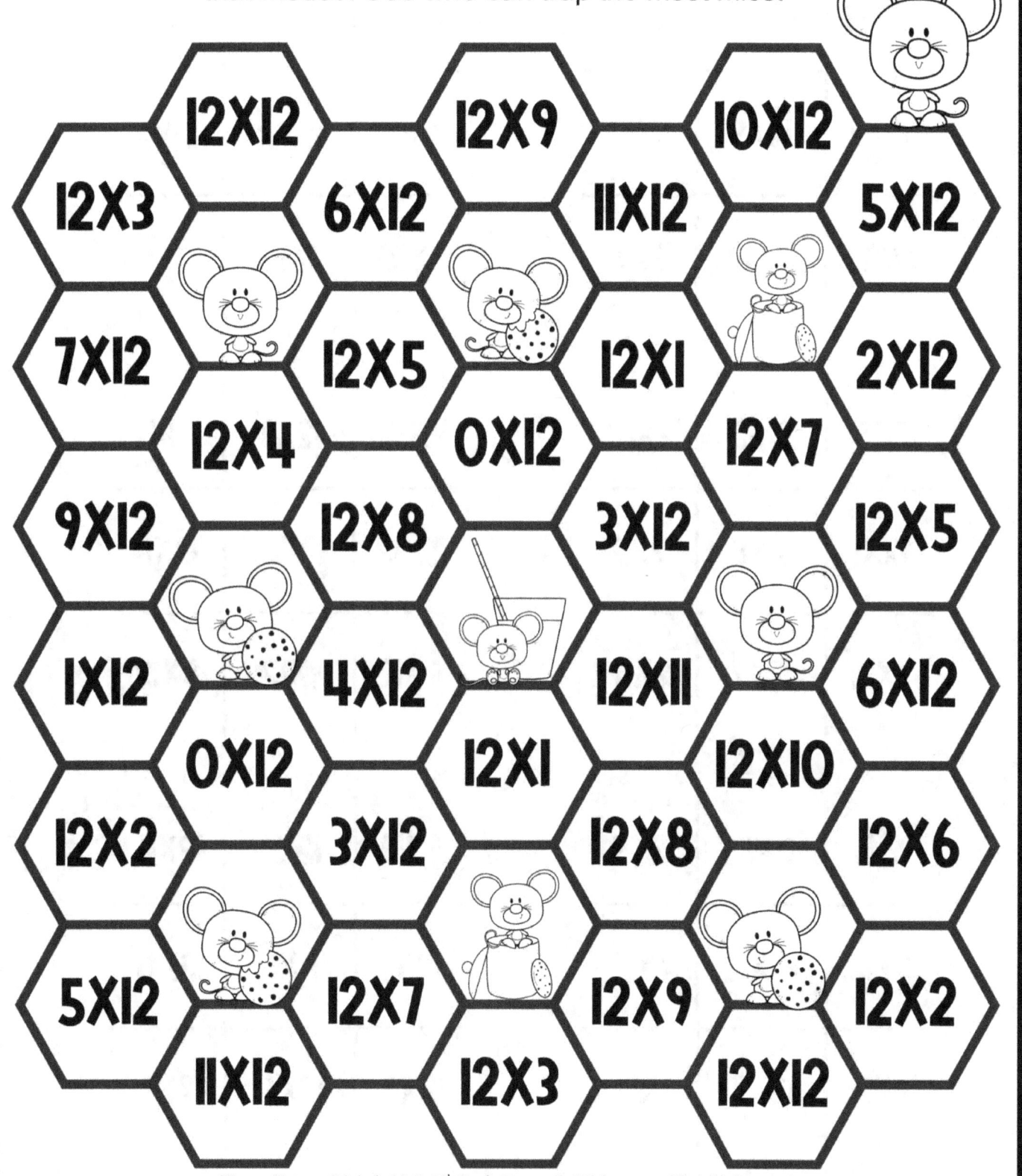

TIC-TAC-TOE: MULTIPLY BY 12

DIRECTIONS: Play a game of tic-tac-toe! Before you mark a space as yours, you must solve the problem in that space.

0X12	12X12	12X1
12X7	12X9	11X12
12X10	8X12	12X5

12X6	7X12	2X12
0X12	12X8	12X3
9X12	1X12	12X4

12X9	5X12	12X10
12X1	12X8	12X2
3X12	12X7	12X12

4X12	12X3	12X12
5X12	12X12	12X11
7X12	1X12	8X12

12X7	0X12	12X5
6X12	10X12	4X12
12X2	12X11	3X12

6X12	12X3	12X4
11X12	10X12	12X0
12X9	12X6	12X7

SPIN A PROBLEM: MULTIPLY BY 12
YOU NEED: PAPERCLIP ✏ PENCIL ✏ CRAYONS

DIRECTIONS: Use a paperclip and pencil to make a spinner. On your turn, spin the paperclip. Multiply that number by 12 and color the product in the table. Each player uses a different color. If the product is not open, your turn is over. See who can solve the most problems!

12	84	144	48	24	96
48	108	72	60	120	36
132	24	36	96	12	48
72	144	60	120	108	132
60	36	108	24	84	72

PIG IN A PEN: MULTIPLY BY 12
YOU NEED: PAPERCLIP 📎 PENCIL ✏️

DIRECTIONS: On their turn, each player spins a number. Multiply the number by 12 and say the answer. If the answer is correct, draw a line to connect 2 dots. When a player completes a box, they write their initial in the box. At the end of the game, boxes are worth 1 point, and boxes with a pig in them are worth 5 points!

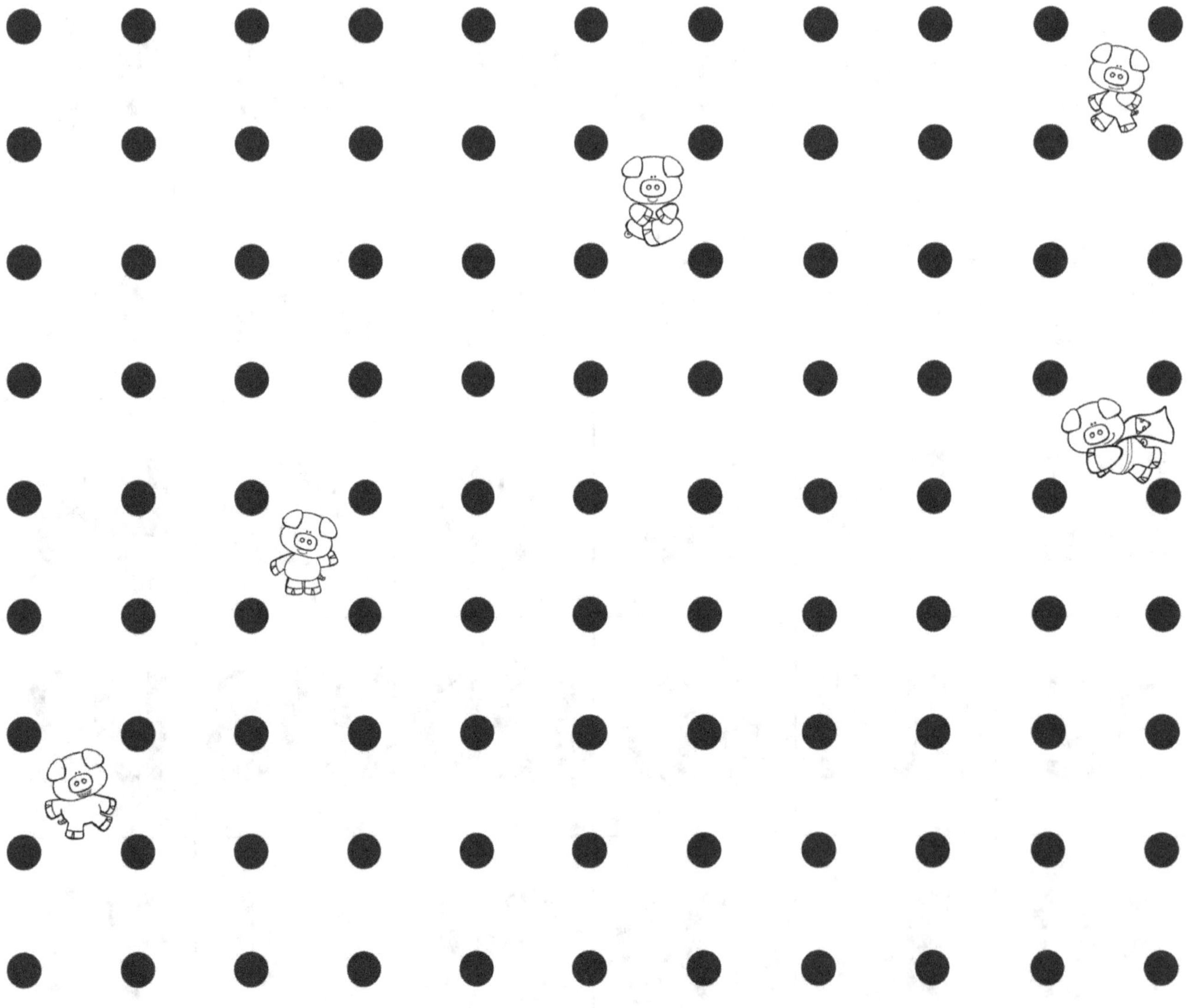

MULTIPLES OF 12 HUNT
YOU NEED: CRAYONS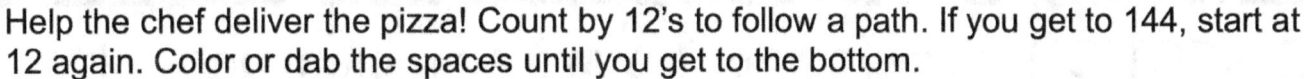

Help the chef deliver the pizza! Count by 12's to follow a path. If you get to 144, start at 12 again. Color or dab the spaces until you get to the bottom.

				START →	12	24	30	48	21	38	56	42
					20	14	36	60	84	81	92	99
					36	27	23	32	72	84	90	108
					6	45	56	63	100	34	96	110
22	48	80	60	48	44	54	64	112	120	108	19
63	56	40	72	109	36	72	35	144	132	29	27
64	72	84	9	108	99	24	12	56	45	36	29
21	96	18	7	24	77	90	63	54	80	72	38
108	27	34	30	48	36	81	72				
8	120	45	24	72	81	45	18				
132	144	12	63	36	92	60	74				
78	40	48	35	71	48	112	72	END →			

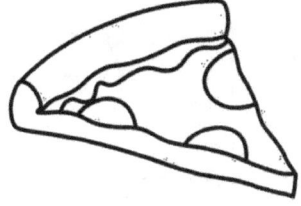

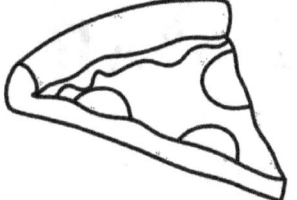

FOUR PROBLEMS IN A ROW: MULTIPLY BY 12
YOU NEED: CRAYONS OR COUNTERS

DIRECTIONS: On your turn, solve a problem and color it or cover it. Each player uses a different color. The first player to get 4 in a row wins!

12 X 12	5 X 12	12 X 9	3 X 12	10 X 12	4 X 12
2 X 12	0 X 12	12 X 9	6 X 12	12 X 1	12 X 7
12 X 5	12 X 10	12 X 12	12 X 4	8 X 12	12 X 2
7 X 12	12 X 8	12 X 0	3 X 12	4 X 12	1 X 12
6 X 12	11 X 12	12 X 7	12 X 6	1 X 12	12 X 11
12 X 2	5 X 12	9 X 12	8 X 12	12 X 10	12 X 3

©Laura Putman Digitals LLC, 3rd Grade Engaged, 2024-present All rights reserved.

FOOTBALL FACTS X12'S

YOU NEED: 1 DIE COUNTERS

DIRECTIONS: On your turn, roll a die. Move that number of spaces and solve the problem on the space. If your answer is incorrect, go back to where you started. The first player to the end, wins!

START

| 5X12 | 6X12 | 12X4 | 7X12 | 12X12 | 12X0 | 2X12 | 12X9 |

8X12

| 2X12 | 12X4 | 6X12 | 12X8 | 3X12 | 12X1 | | 12X3 |

| 12X10 | | | | | 12X10 | | 10X12 |

END

| 1X12 | | 7X12 | | | 4X12 | | 1X12 |

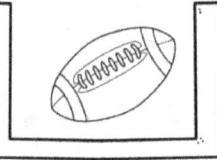

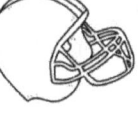

| 12X2 | | 12X9 | 12X0 | 11X12 | 5X12 | | 11X12 |

| 12X11 | | | | | | | 5X12 |

| 8X12 | 12X7 | 12X12 | 0X12 | 12X6 | 12X9 | 12X3 | 4X12 |

ROLL & SOLVE: MIXED FACTS
YOU NEED: 1 DIE, CRAYONS

DIRECTIONS: Assign one player even numbers on the die and the other player odd numbers. Take turns rolling. If a player rolls one of their numbers on the die, they solve the next problem under that die and color the space. If they do not roll one of their numbers, their turn is over. See who can fill their columns first!

⚀	⚁	⚂	⚃	⚄	⚅
4X4	0X5	7X3	5X5	8X9	8X3
12X6	7X7	6X10	2X8	4X5	0X8
5X6	12X3	3X4	3X3	12X10	6X3
7X5	1X4	0X11	4X6	2X7	5X9
9X9	9X10	8X8	8X11	2X2	2X4
0X10	8X5	12X1	9X3	7X6	11X10

TRAP THE MONSTERS: MIXED FACTS
YOU NEED: COUNTERS

DIRECTIONS: On your turn, solve a problem. Put a counter on the problem. The last player who puts a counter around a monster traps it! Put a counter on that monster. See who can trap the most monsters.

- 12X1, 9X9, 4X10
- 3X0, 6X0, 11X12, 5X3
- 6X3, 12X10, 5X2, 11X12
- 10X10, 10X9, 7X8
- 9X6, 4X8, 1X3, 5X7
- 2X2, 7X4, 5X11, 6X8
- 0X0, 7X7, 4X4
- 5X9, 2X4, 8X8, 4X6
- 6X5, 6X7, 8X9, 2X3
- 1X0, 3X4, 5X5

MIXED FACTS TIC-TAC-TOE

DIRECTIONS: Play a game of tic-tac-toe! Before you mark a space as yours, you must solve the problem in that space.

6X3	8X8	12X4		6X6	7X10	11X2
2X5	7X9	11X11		0X12	5X6	3X8
1X10	9X8	5X4		9X10	7X2	3X4

8X6	5X5	12X11		9X2	3X3	0X0
4X9	5X8	6X10		5X12	9X3	4X11
2X3	4X4	1X2		7X7	6X2	8X1

7X5	8X2	3X5		8X2	1X3	4X2
6X6	10X10	9X5		7X4	5X10	12X9
4X8	12X12	3X7		11X5	6X4	9X9

SPIN A PROBLEM: MIXED FACTS
YOU NEED: PAPERCLIP ✐ PENCIL ✏ CRAYONS

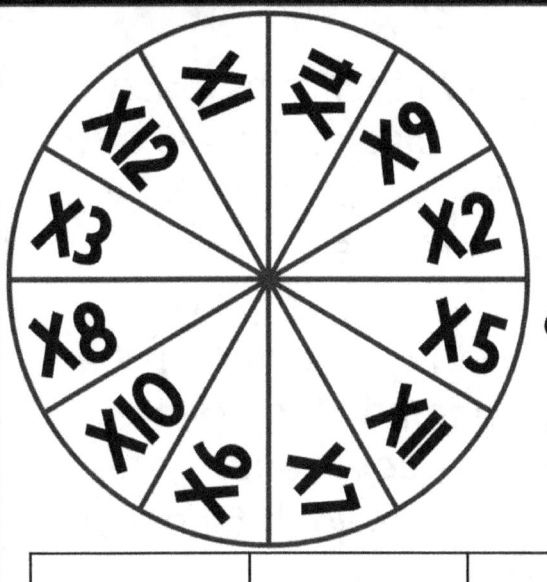

DIRECTIONS: Use a paperclip and pencil to make a spinner. On your turn, spin the paperclip. Multiply that number by a number of your choice in the grid and color it. Each player uses a different color. See who can solve the most problems!

1	7	12	4	2	8
4	9	6	5	10	3
11	2	0	8	1	4
6	12	5	10	9	11
0	3	9	2	7	6

PIG IN A PEN: MIXED FACTS
YOU NEED: PAPERCLIP / PENCIL / 2 DICE

DIRECTIONS: On their turn, each player rolls both dice and spins a number. Multiply the numbers. If the answer is correct, draw a line to connect 2 dots. When a player completes a box, they write their initial in the box. At the end of the game, boxes are worth 1 point, and boxes with a pig in them are worth 5 points!

ARRAYS? HOORAY! MIXED FACTS
YOU NEED: CRAYONS   2 DICE

DIRECTIONS: On their turn, each player rolls both dice twice. Make an array using the number you rolled. Each player uses a different color to make their arrays. Write the multiplication problem inside the array. If your array won't fit, your turn is over. When no more arrays can be made, the game is over. Whoever makes the most arrays, wins!

MULTIPLICATION FACTS PUZZLE
YOU NEED: GLUE SCISSORS

DIRECTIONS: Cut the puzzle pieces from the next page apart. To solve the puzzle, glue the piece with the matching product on each problem. Do not cut this page apart.

1x2	8x5	7x4	9x6	2x2
2x9	1x1	6x5	0x3	4x8
7x10	4x4	5x9	3x1	11x10
6x7	3x11	8x8	3x2	6x8
10x12	2x4	4x9	9x11	5x7

©Laura Putman Digitals LLC, 3rd Grade Engaged, 2024-present All rights reserved.

MULTIPLICATION FACTS PUZZLE

DIRECTIONS: Cut these puzzle pieces apart. Glue the piece with the correct product on top of each matching problem on the previous page.

©Laura Putman Digitals LLC, 3rd Grade Engaged, 2024-present All rights reserved.

FOUR PROBLEMS IN A ROW: MIXED FACTS
YOU NEED: CRAYONS OR COUNTERS

DIRECTIONS: On your turn, solve a problem and color it or cover it. Each player uses a different color. The first player to get 4 in a row wins!

2 X 2	0 X 8	9 X 9	6 X 4	2 X 1	5 X 7
5 X 3	10 X 10	12 X 3	4 X 9	8 X 8	2 X 8
7 X 7	4 X 8	0 X 0	9 X 3	5 X 4	1 X 3
6 X 5	12 X 11	3 X 7	6 X 6	1 X 1	11 X 2
4 X 4	5 X 5	7 X 6	5 X 9	10 X 8	3 X 2
10 X 12	3 X 4	6 X 8	9 X 2	2 X 5	4 X 1

POPPIN' PRODUCTS: MIXED FACTS

YOU NEED: 1 DIE COUNTERS

DIRECTIONS: On your turn, roll a die. Move that number of spaces and solve the problem on the space. If your answer is incorrect, go back to where you started. The first player to the end, wins!

START

| 5X3 | 6X8 | 4X4 | 7X4 | 12X10 | 0X0 | 2X3 | 8X9 |

7X7

| 10X10 | 2X4 | 6X3 | 7X8 | 2X2 | 12X1 | | 4X8 |

| 8X8 | | | | | 4X10 | | 10X9 |

END

| 6X0 | | 6X7 | | | 5X7 | | 1X3 |

| 5X2 | | 3X4 | 3X0 | 11X12 | 4X3 | | 1X0 |

| 11X11 | | | | | | | 5X11 |

| 8X8 | 3X7 | 9X9 | 6X5 | 9X6 | 5X9 | 5X5 | 4X6 |

ROLL & SOLVE: MULTIPLES OF 10
YOU NEED: 1 DIE, CRAYONS

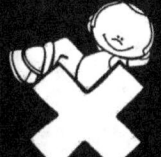

DIRECTIONS: Assign one player even numbers on the die and the other player odd numbers. Take turns rolling. If a player rolls one of their numbers on the die, they solve the next problem under that die and color the space. If they do not roll one of their numbers, their turn is over. See who can fill their columns first!

⚀	⚁	⚂	⚃	⚄	⚅
60X3	80X12	4X50	80X11	7X70	11X60
11X90	20X2	5X20	50X5	80X4	70X5
80X7	4X70	90X6	0X50	6X30	1X40
8X80	110X10	12X90	40X12	50X6	0X70
40X10	10X0	20X4	30X10	11X70	10X10
20X10	70X9	12X10	8X30	20X7	40X6

©Laura Putman Digitals LLC, 3rd Grade Engaged, 2024-present All rights reserved.

TRAP THE POTATOES: MULTIPLES OF 10
YOU NEED: COUNTERS

DIRECTIONS: On your turn, solve a problem. Put a counter on the problem. The last player who puts a counter around a potato traps it! Put a counter on that potato. See who can trap the most potatoes.

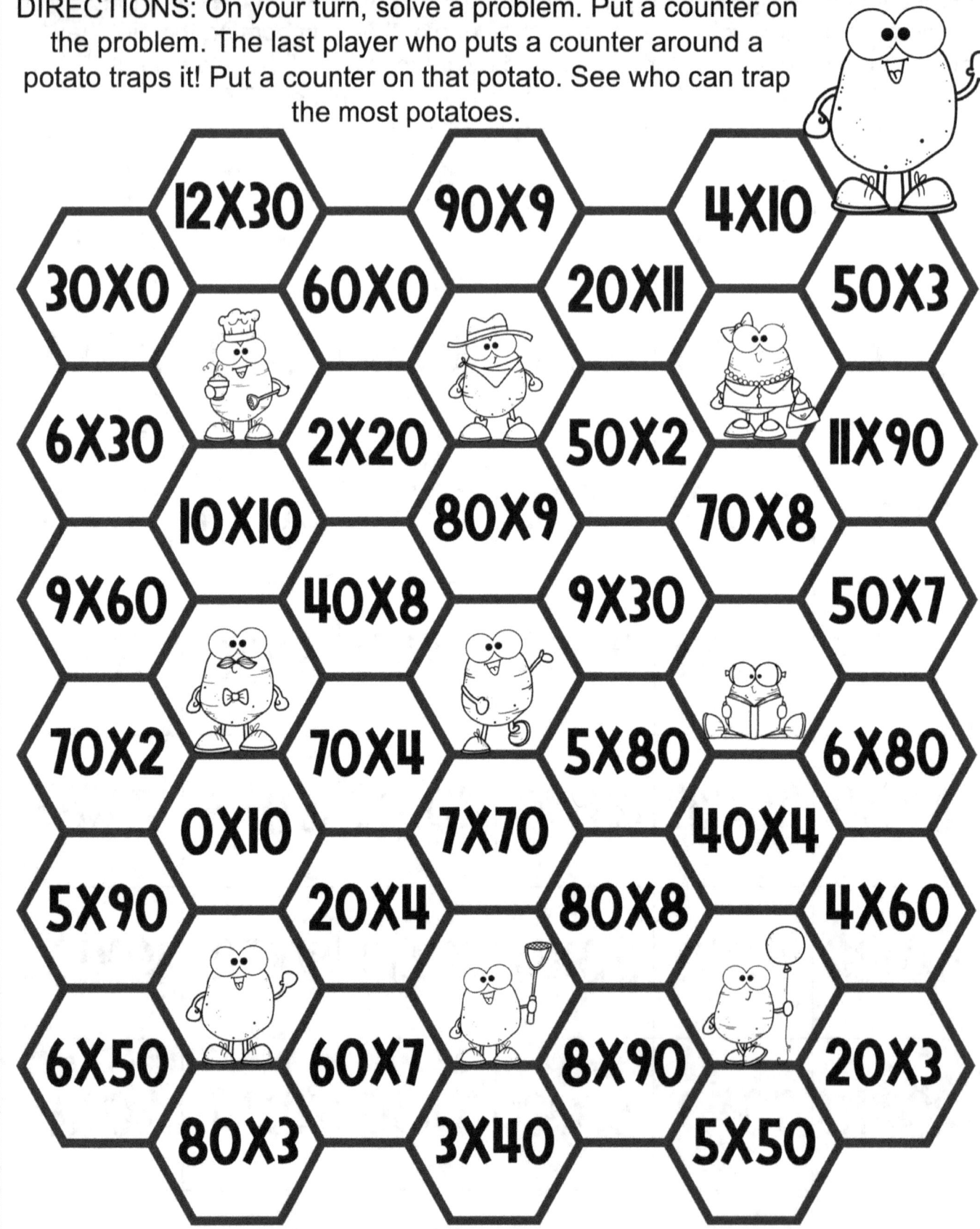

TIC-TAC-TOE: MULTIPLES OF 10

DIRECTIONS: Play a game of tic-tac-toe! Before you mark a space as yours, you must solve the problem in that space.

10X12	60X5	40X1
90X7	30X9	11X20
10X10	8X80	40X5

60X6	7X30	2X80
0X10	40X8	50X3
9X90	70X12	60X4

20X9	5X50	12X80
20X1	70X8	12X30
3X80	60X7	40X11

4X40	30X3	10X12
5X20	60X12	30X11
7X20	1X30	8X90

50X7	0X40	80X5
6X30	10X11	4X20
30X2	90X11	3X40

60X0	10X3	70X4
11X80	10X40	50X0
60X9	20X6	10X7

MULTIPLES OF 10 PUZZLE
YOU NEED: GLUE SCISSORS

DIRECTIONS: Cut the puzzle pieces from the next page apart. To solve the puzzle, glue the piece with the matching product on each problem. Do not cut this page apart.

4 x 60	30 x 7	10 x 12	2 x 20
50 x 11	3 x 60	10 x 40	9 x 30
9 x 90	2 x 80	50 x 12	80 x 8
70 x 8	90 x 10	4 x 80	4 x 90
40 x 7	2 x 30	4 x 50	50 x 11

MULTIPLES OF 10 PUZZLE

DIRECTIONS: Cut these puzzle pieces apart. Glue the piece with the correct product on top of each matching problem on the previous page.

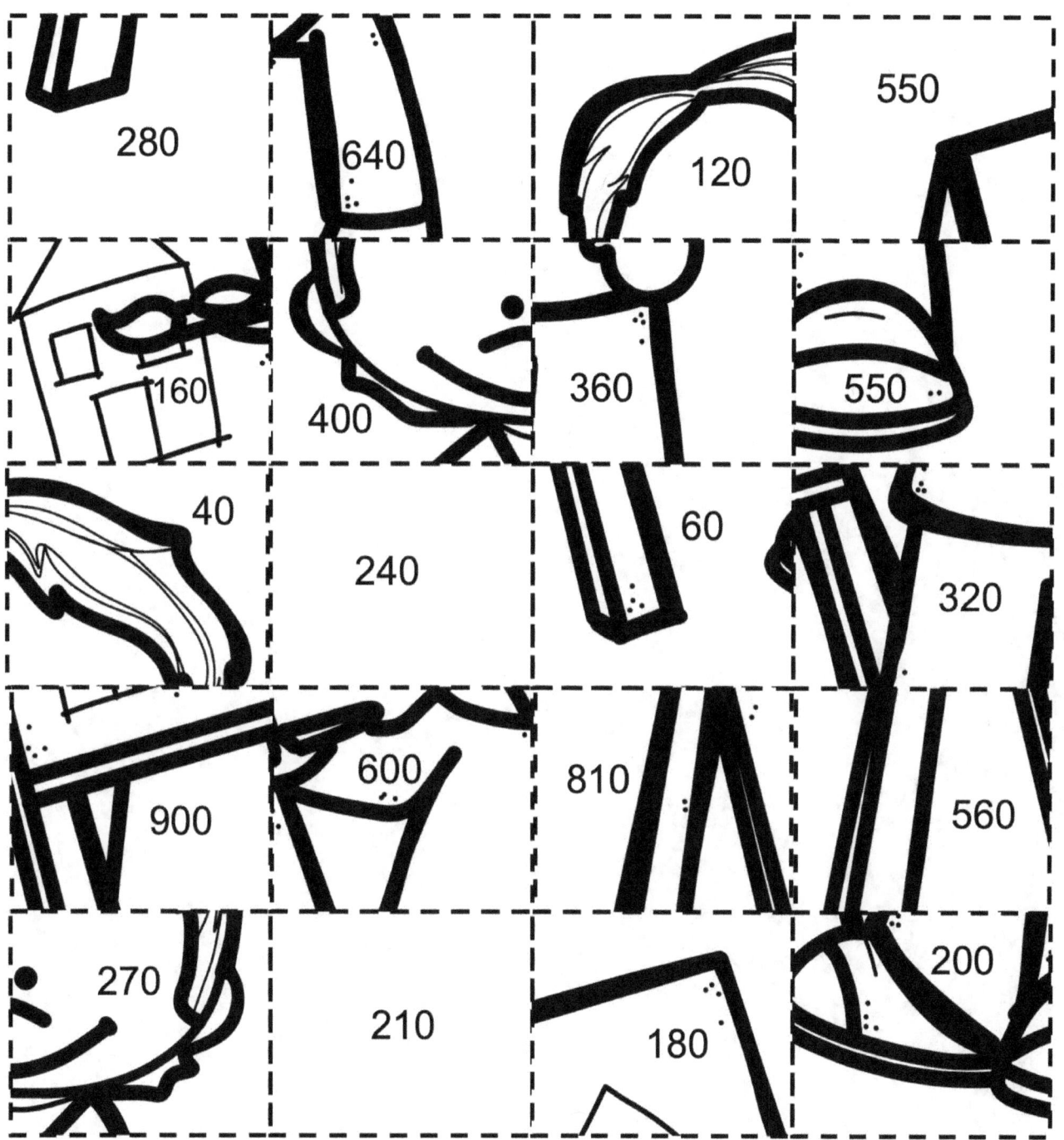

280	640	120	550
160	400	360	550
40	240	60	320
900	600	810	560
270	210	180	200

FOUR PROBLEMS IN A ROW: MULTIPLES OF 10
YOU NEED: CRAYONS OR COUNTERS

DIRECTIONS: On your turn, solve a problem and color it or cover it. Each player uses a different color. The first player to get 4 in a row wins!

30 X 12	5 X 80	90 X 9	3 X 40	10 X 70	4 X 60
20 X 11	0 X 10	70 X 9	6 X 50	80 X 1	12 X 70
50 X 5	12 X 20	4 X 40	50 X 4	8 X 90	30 X 2
7 X 60	30 X 8	30 X 0	3 X 50	4 X 20	1 X 60
6 X 60	11 X 90	50 X 7	20 X 6	1 X 10	80 X 11
70 X 2	5 X 90	9 X 40	8 X 40	10 X 10	30 X 3

©Laura Putman Digitals LLC, 3rd Grade Engaged, 2024-present All rights reserved.

POPPIN' PRODUCTS: MULTIPLES OF 10

YOU NEED: 1 DIE 🎲 COUNTERS ⚫🔴

DIRECTIONS: On your turn, roll a die. Move that number of spaces and solve the problem on the space. If your answer is incorrect, go back to where you started. The first player to the end, wins!

START

| 50X3 | 6X80 | 40X4 | 70X4 | 12X10 | 10X0 | 20X3 | 8X90 |

| | | | | | | | 70X7 |

| 10X10 | 2X40 | 60X3 | 70X8 | 2X02 | 12X10 | | 4X80 |

| 8X80 | | **END** | | | 40X10 | | 10X90 |

| 60X0 | | 6X70 | | | 5X70 | | 1X30 |

| 50X2 | | 3X40 | 30X0 | 50X12 | 4X30 | | 10X0 |

| 11X20 | | | | | | | 50X11 |

| 8X80 | 30X7 | 90X9 | 6X50 | 90X6 | 5X90 | 50X5 | 4X60 |

ROLL & SOLVE: ASSOCIATIVE PROPERTY
YOU NEED: 1 DIE CRAYONS

DIRECTIONS: Assign one player even numbers on the die and the other player odd numbers. Take turns rolling. If a player rolls one of their numbers on the die, they solve the next problem under that die and color the space. If they do not roll one of their numbers, their turn is over. See who can fill their columns first!

⚀	⚁	⚂	⚃	⚄	⚅
(2X2)X6	5X(2X3)	(3X3)X2	3X(3X4)	(2X5)X7	8X(2X4)
11X(3X3)	(2X5)X10	7X(4X2)	(2x2)x2	2x(3x1)	6x(9x0)
(6x2)x1	5x(1x5)	(2x2)x11	9x(3x3)	(2x4)x5	2x(5x2)
11x(4x2)	(4x3)x12	7x(3x3)	(2x2)x7	5x(2x2)	(3x2)x3
(4x3)x9	5x(3x3)	(3x0)x3	1x(1x1)	(5x2)x11	8x(3x2)
7x(3x2)	(2x2)x3	11x(1x5)	(4x2)x2	7x(4x3)	(5x2)x8

TRAP THE CUPCAKES: ASSOCIATIVE PROPERTY
YOU NEED: COUNTERS

DIRECTIONS: On your turn, solve a problem. Put a counter on the problem. The last player who puts a counter around a cupcake traps it! Put a counter on that cupcake. See who can trap the most cupcakes.

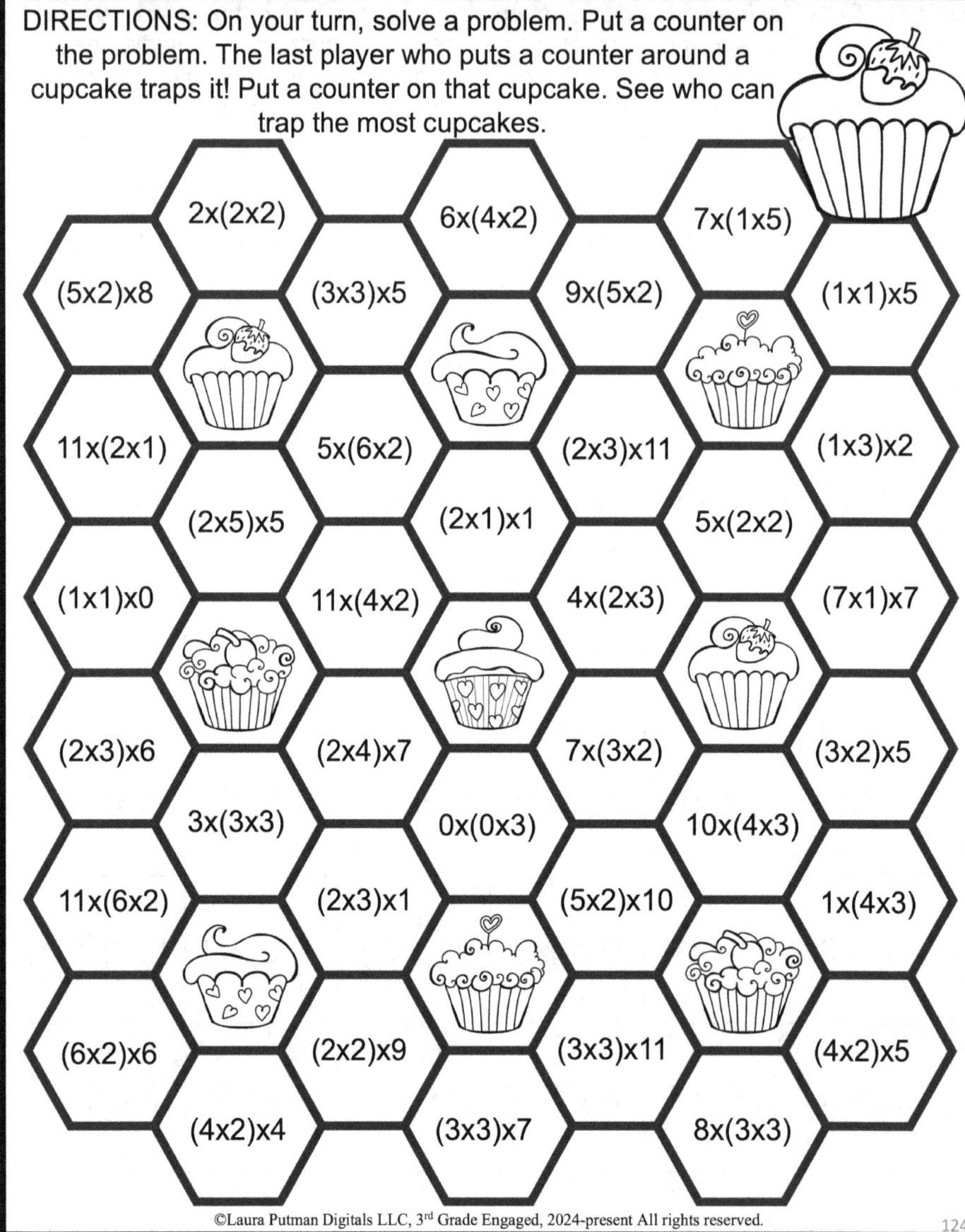

Hexagon problems:
- 2x(2x2)
- 6x(4x2)
- 7x(1x5)
- (5x2)x8
- (3x3)x5
- 9x(5x2)
- (1x1)x5
- 11x(2x1)
- 5x(6x2)
- (2x3)x11
- (1x3)x2
- (2x5)x5
- (2x1)x1
- 5x(2x2)
- (1x1)x0
- 11x(4x2)
- 4x(2x3)
- (7x1)x7
- (2x3)x6
- (2x4)x7
- 7x(3x2)
- (3x2)x5
- 3x(3x3)
- 0x(0x3)
- 10x(4x3)
- 11x(6x2)
- (2x3)x1
- (5x2)x10
- 1x(4x3)
- (6x2)x6
- (2x2)x9
- (3x3)x11
- (4x2)x5
- (4x2)x4
- (3x3)x7
- 8x(3x3)

©Laura Putman Digitals LLC, 3rd Grade Engaged, 2024-present All rights reserved.

ASSOCIATIVE PROPERTY TIC-TAC-TOE

DIRECTIONS: Play a game of tic-tac-toe! Before you mark a space as yours, you must solve the problem in that space.

(2x3)x6	5x(2x2)	(3x3)x7		(2x2)x4	11x(6x2)	(5x2)x8
10x(4x3)	11x(4x2)	(1x3)x2		3x(2x3)	7x(1x5)	(1x1)x0
(2x4)x7	(3x2)x5	9x(3x3)		(3x3)x11	(4x2)x3	3x(4x3)

(7x1)x7	9x(5x2)	(2x1)x1		(2x2)x11	7x(2x2)	(6x2)x6
4x(2x3)	7x(3x2)	(4x3)x2		3x(11x1)	3x(3x3)	(4x2)x4
(2x0)x5	(4x2)x5	8x(3x3)		(3x3)x5	(1x1)x5	2x(2x5)

(5x1)x5	5x(6x2)	(2x2)x9		(2x5)x5	11x(2x1)	(2x3)x1
0x(0x3)	6x(4x2)	(2x3)x11		1x(4x3)	0x(4x2)	(1x3)x5
(5x2)x10	(4x3)x4	7x(1x3)		(3x4)x7	(3x2)x2	2x(2x2)

ASSOCIATIVE PROPERTY PUZZLE
YOU NEED: GLUE SCISSORS

DIRECTIONS: Cut the puzzle pieces from the next page apart. To solve the puzzle, glue the piece with the matching product on each problem. Do not cut this page apart.

6 x (2 x 2)	(3 x 3) x 2	(4 x 2) x 8	9 x (2 x 3)
3 x (2 x 1)	(3 x 2) x 5	7 x (2 x 3)	(4 x 1) x 7
(5 x 2) x 5	6 x (2 x 3)	0 x (1 x 1)	(2 x 2) x 5
3 x (3 x 3)	1 x (1 x 1)	(4 x 2) x 4	(3 x 3) x 7

ASSOCIATIVE PROPERTY PUZZLE

DIRECTIONS: Cut these puzzle pieces apart. Glue the piece with the correct product on top of each matching problem on the previous page.

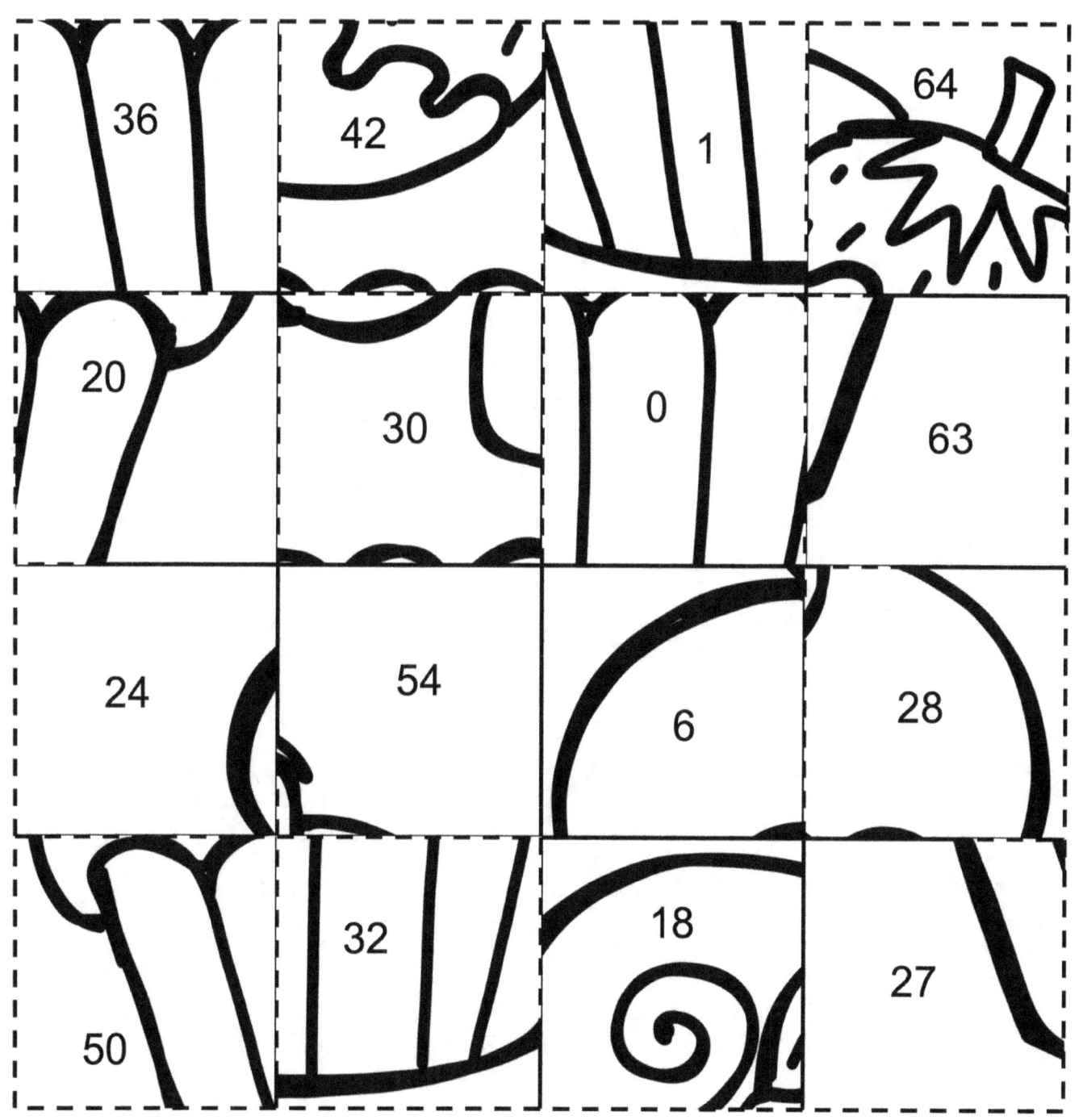

FOUR PROBLEMS IN A ROW: ASSOCIATIVE PROPERTY
YOU NEED: CRAYONS OR COUNTERS

DIRECTIONS: On your turn, solve a problem and color it or cover it. Each player uses a different color. The first player to get 4 in a row wins!

(4x2)x6	7x(2x2)	(3x3)x3	11x(2x5)	(5x1)x3	1x(4x2)
5x(6x2)	(3x3)x7	2x(3x2)	(5x2)x12	3x(4x2)	(2x2)x2
(2x8)x0	11x(3x3)	(2x4)x8	7x(2x3)	(2x2)x5	7x(1x7)
3x(2x2)	(2x3)x5	4x(3x3)	(4x3)x11	3x(2x1)	(1x0)x1
(5x2)x9	7x(3x3)	(3x2)x6	5x(4x2)	(2x2)x6	2x(3x3)
7x(5x2)	(3x2)x3	5x(3x3)	3x(3x3)	(2x5)x10	2x(6x2)

MULTIPLICATION MYSTERY: ASSOCIATIVE PROPERTY

YOU NEED: 1 DIE 🎲 COUNTERS ⚫⚫

DIRECTIONS: On your turn, roll a die. Move that number of spaces and solve the problem on the space. If your answer is incorrect, go back to where you started. The first player to the end, wins!

START

(4x2)x4		(2x1)x1	(2x4)x7	7x(3x2)	(5x2)x8	4x(2x3) **END**
(2x3)x1		(1x1)x5		11x(4x2)	(2x3)x6
11x(2x1)		10x(4x3)		3x(3x3)	(7x1)x7
(3x3)x7		6x(4x2)		9x(5x2)	(3x3)x11
0x(0x3)		(2x5)x5		11x(6x2)	2x(2x2)
(4x2)x5		(5x2)x10		(6x2)x6	(1x3)x2
8x(3x3)		(2x3)x11		(1x1)x0	(3x3)x5
5x(2x2)	1x(4x3)	(2x2)x9	(3x2)x5	5x(6x2)	7x(1x5)

TRAP THE STARS: COMMUTATIVE PROPERTY
YOU NEED: COUNTERS

DIRECTIONS: On your turn, choose a problem and name a related multiplication fact. Put a counter on the problem. The last player who puts a counter around a star traps it! Put a counter on that star. See who can trap the most stars.

8X9	9X9	4X10
4X6	2X2	6X5	7X4

6X3	12X10	5X2	11X12
6X8	11X11	3X0
9X6	4X8	1X3	5X7

5X3	12X1	5X11	6X0
0X0	7X7	4X4
5X9	2X4	8X8	10X10

| 7X8 | 6X7 | 10X9 | 2X3 |
| 1X0 | 3X4 | 5X5 |

©Laura Putman Digitals LLC, 3rd Grade Engaged, 2024-present All rights reserved.

COMMUTATIVE PROPERTY PUZZLE
YOU NEED: GLUE SCISSORS

DIRECTIONS: Cut the puzzle pieces from the next page apart. To solve the puzzle, glue the piece with the missing number on each set of problems. Do not cut this page apart.

? x 1 = 9 1 x ? = 9	? x 8 = 24 8 x ? = 24	6 x 7 = ? 7 x 6 = ?
3 x ? = 33 ? x 3 = 33	? x 12 = 48 12 x ? = 48	5 x ? = 0 ? x 5 = 0
? x 4 = 8 4 x ? = 8	7 x 10 = ? 10 x 7 = ?	9 x ? = 72 ? x 9 = 72
3 x 6 = ? 6 x 3 = ?	3 x 5 = ? 5 x 3 = ?	? x 5 = 50 5 x ? = 50
4 x 11 = ? 11 x 4 = ?	8 x 4 = ? 4 x 8 = ?	5 x ? = 5 ? x 5 = 5

COMMUTATIVE PROPERTY PUZZLE

DIRECTIONS: Cut these puzzle pieces apart. Glue the piece with the correct missing number on top of each matching set of problems on the previous page.

FOUR PROBLEMS IN A ROW: COMMUTATIVE PROPERTY
YOU NEED: CRAYONS OR COUNTERS

DIRECTIONS: On your turn, choose a problem. Name a related multiplication fact and color it or cover it. Each player uses a different color. The first player to get 4 in a row wins!

6 X 12	5 X 5	8 X 9	3 X 3	11 X 10	7 X 4
2 X 2	0 X 8	9 X 9	6 X 4	2 X 1	5 X 7
5 X 3	10 X 10	12 X 3	4 X 9	8 X 8	2 X 8
7 X 7	4 X 8	0 X 0	9 X 3	5 X 4	1 X 3
6 X 5	12 X 11	3 X 7	6 X 6	1 X 1	11 X 2
4 X 4	5 X 5	7 X 6	5 X 9	10 X 8	3 X 2

FOOTBALL FACTS: COMMUTATIVE PROPERTY

YOU NEED: 1 DIE 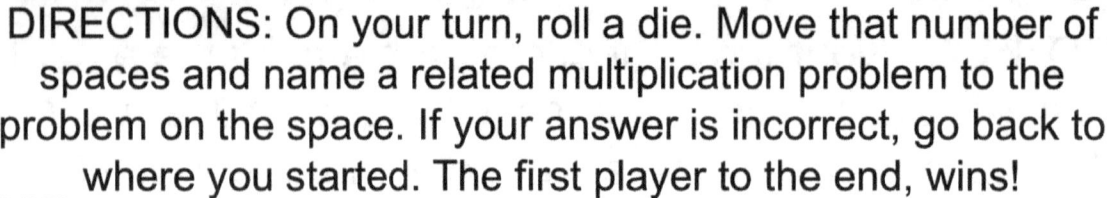 COUNTERS

DIRECTIONS: On your turn, roll a die. Move that number of spaces and name a related multiplication problem to the problem on the space. If your answer is incorrect, go back to where you started. The first player to the end, wins!

START

| 4X6 | 5X2 | 4X4 | 7X4 | 12X10 | 0X0 | 5X3 | 8X9 |

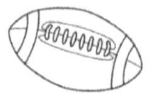    7X7

| 10X10 | 2X4 | 6X3 | 7X8 | 2X2 | 12X1 | | 4X8 |

8X8 | | | | | 4X10 | | 10X9

 **END**

6X0 | | 1X3 | | | 5X7 | | 6X7

| 2X3 |  | 3X4 | 3X0 | 11X12 | 4X3 |  | 1X0 |

| 11X11 | | | | | | | 5X11 |

| 8X8 | 3X7 | 9X9 | 6X5 | 9X6 | 5X9 | 5X5 | 6X8 |

FOUR PROBLEMS IN A ROW: DISTRIBUTIVE PROPERTY
YOU NEED: CRAYONS OR COUNTERS

DIRECTIONS: On your turn, solve a problem using the Distributive Property and color it or cover it. Each player uses a different color. The first player to get 4 in a row wins!

9 x 9 __x__ + __x__ ___ + ___ = ___	4 x 6 __x__ + __x__ ___ + ___ = ___	3 x 12 __x__ + __x__ ___ + ___ = ___	5 x 7 __x__ + __x__ ___ + ___ = ___
6 x 8 __x__ + __x__ ___ + ___ = ___	3 x 11 __x__ + __x__ ___ + ___ = ___	2 x 10 __x__ + __x__ ___ + ___ = ___	4 x 9 __x__ + __x__ ___ + ___ = ___
8 x 8 __x__ + __x__ ___ + ___ = ___	7 x 4 __x__ + __x__ ___ + ___ = ___	5 x 11 __x__ + __x__ ___ + ___ = ___	10 x 12 __x__ + __x__ ___ + ___ = ___
11 x 12 __x__ + __x__ ___ + ___ = ___	5 x 6 __x__ + __x__ ___ + ___ = ___	9 x 3 __x__ + __x__ ___ + ___ = ___	7 x 7 __x__ + __x__ ___ + ___ = ___
8 x 7 __x__ + __x__ ___ + ___ = ___	12 x 12 __x__ + __x__ ___ + ___ = ___	8 x 9 __x__ + __x__ ___ + ___ = ___	4 x 6 __x__ + __x__ ___ + ___ = ___

DISTRIBUTIVE PROPERTY PUZZLE
YOU NEED: GLUE SCISSORS

DIRECTIONS: Cut the puzzle pieces from the next page apart. To solve the puzzle, glue the piece with the product on each problem. Do not cut this page apart.

9 x 9	7 x 3	10 x 7
___x___ + ___x___	___x___ + ___x___	___x___ + ___x___
___ + ___	___ + ___	___ + ___
= ___	= ___	= ___
12 x 6	8 x 5	11 x 4
___x___ + ___x___	___x___ + ___x___	___x___ + ___x___
___ + ___	___ + ___	___ + ___
= ___	= ___	= ___
3 x 9	8 x 8	11 x 11
___x___ + ___x___	___x___ + ___x___	___x___ + ___x___
___ + ___	___ + ___	___ + ___
= ___	= ___	= ___
10 x 6	9 x 2	12 x 7
___x___ + ___x___	___x___ + ___x___	___x___ + ___x___
___ + ___	___ + ___	___ + ___
= ___	= ___	= ___

©Laura Putman Digitals LLC, 3rd Grade Engaged, 2024-present All rights reserved.

DISTRIBUTIVE PROPERTY PUZZLE

DIRECTIONS: Cut these puzzle pieces apart. Glue the piece with the correct product on top of each matching problem on the previous page.

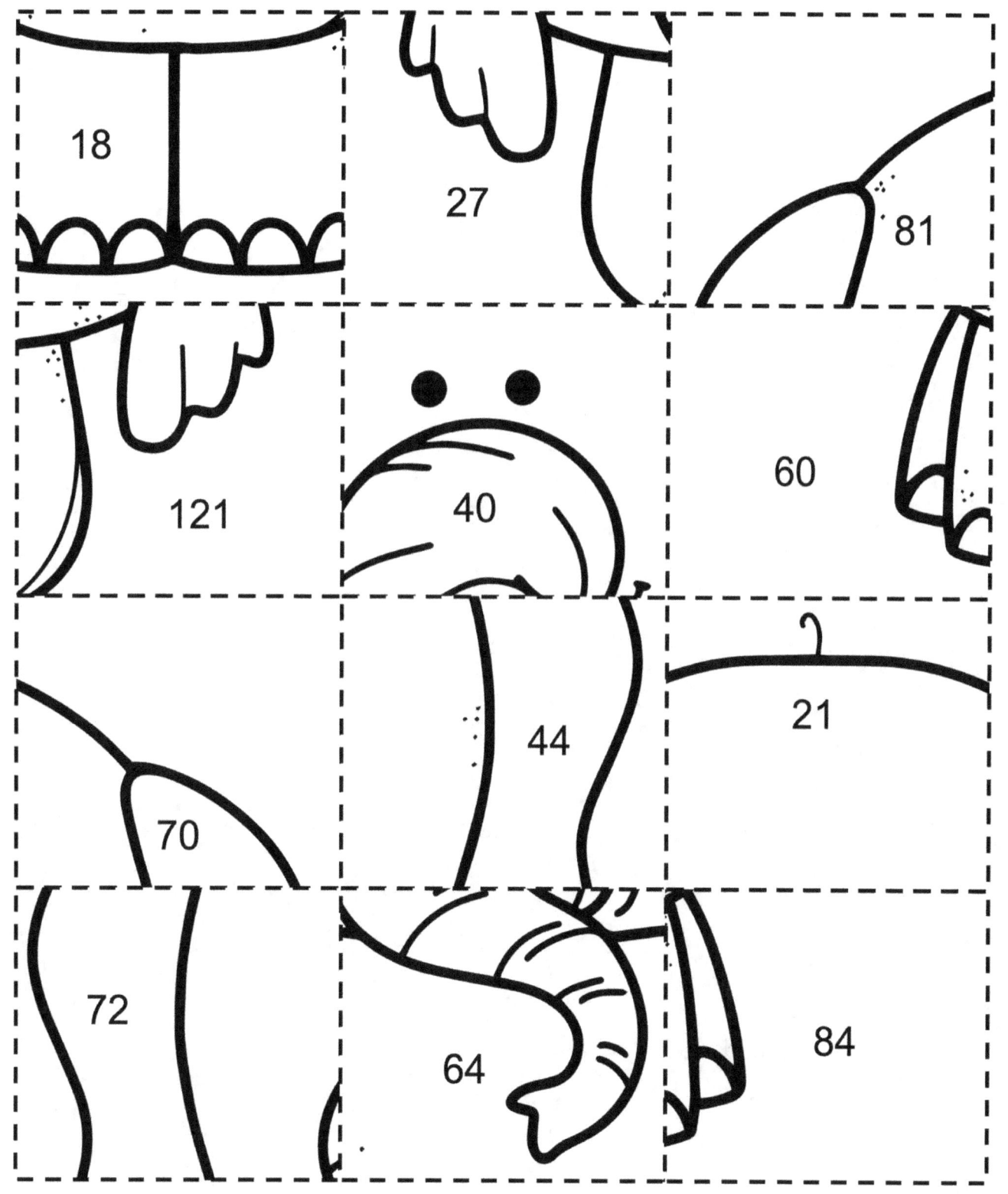

FOUR PROBLEMS IN A ROW: REPEATED ADDITION
YOU NEED: CRAYONS OR COUNTERS

DIRECTIONS: On your turn, choose a problem. Say the multiplication sentence that it equals and color it or cover it. Each player uses a different color. The first player to get 4 in a row wins!

1 + 1 + 1 + 1 + 1 + 1 + 1	4 + 4 + 4 + 4 + 4 + 4 + 4 + 4 + 4	6 + 6 + 6 + 6 + 6 + 6	3 + 3 + 3	2 + 2 + 2 + 2 + 2 + 2 + 2 + 2 + 2	10 + 10 + 10 + 10
7 + 7 + 7 + 7 + 7 + 7 + 7	5 + 5 + 5 + 5 + 5	3 + 3 + 3 + 3 + 3 + 3	11 + 11 + 11 + 11 + 11 + 11 + 11 + 11 + 11	9 + 9 + 9 + 9 + 9 + 9 + 9 + 9 + 9	7x(1x7)
5 + 5 + 5 + 5 + 5 + 5	1 + 1	12 + 12 + 12	10 + 10 + 10 + 10 + 10 + 10	12 + 12	8 + 8 + 8 + 8 + 8 + 8 + 8 + 8 + 8
4 + 4 + 4 + 4	5 + 5 + 5 + 5	3 + 3 + 3 + 3 + 3 + 3 + 3 + 3 + 3 + 3	4 + 4 + 4 + 4 + 4 + 4 + 4	7 + 7 + 7 + 7 + 7	6 + 6 + 6 + 6 + 6 + 6 + 6 + 6 + 6
8 + 8 + 8 + 8 + 8 + 8 + 8 + 8 + 8 + 8	11 + 11	10 + 10 + 10 + 10 + 10	12 + 12 + 12 + 12 + 12	3 + 3 + 3 + 3	9 + 9
0 + 0 + 0 + 0 + 0 + 0 + 0 + 0	11 + 11 + 11 + 11 + 11 + 11 + 11	7 + 7 + 7 + 7 + 7 + 7 + 7	5 + 5 + 5 + 5 + 5 + 5 + 5 + 5 + 5	7 + 7	5 + 5 + 5

TRAP THE BALLOONS: REPEATED ADDITION
YOU NEED: COUNTERS

DIRECTIONS: On your turn, choose a problem and name a related multiplication fact. Put a counter on the problem. The last player who puts a counter around a balloon traps it! Put a counter on that balloon. See who can trap the most balloons.

- 1+1+1+1+1+1+1+1+1+1
- 3+3+3
- 9+9+9+9+9+9+9+9+9
- 10+10+10+10+10
- 4+4+4+4+4+4+4+4
- 5+5
- 12+12+12+12
- 7+7+7+7+7
- 6+6+6+6+6
- 7+7+7+7+7+7+7
- 2+2+2
- 4+4+4+4+4+4
- 9+9+9+9+9
- 3+3+3+3
- 11+11
- 1+1+1+1+1+1
- 4+4+4+4
- 5+5+5+5
- 11+11+11+11+11+11+11
- 5+5+5+5+5
- 6+6+6
- 8+8
- 12+12
- 11+11+11+11
- 4+4+4+4+4+4+4
- 10+10+10+10+10+10
- 7+7+7+7+7+7+7
- 3+3+3+3+3+3+3+3+3+3+3+3
- 1+1+1+1+1+1+1+1+1+1+1
- 9+9+9+9+9
- 8+8+8+8+8+8
- 9+9+9+9+9+9
- 5+5+5
- 10+10+10
- 10+10+10+10
- 12+12+12

TIC-TAC-TOE: REPEATED ADDITION

DIRECTIONS: Play a game of tic-tac-toe! Before you mark a space as yours, you must name the multiplication sentence made by the repeated addition.

6 + 6 + 6 + 6 + 6	7 + 7 + 7 + 7 + 7 + 7	10 + 10 + 10 + 10 + 10
8 + 8 + 8 + 8 + 8 + 8	4 + 4 + 4 + 4 + 4 + 4 + 4 + 4 + 4 + 4 + 4	11 + 11
3 + 3 + 3	3 + 3 + 3 + 3 + 3 + 3 + 3 + 3	2 + 2 + 2 + 2 + 2 + 2 + 2 + 2 + 2
7 + 7 + 7 + 7 + 7 + 7 + 7 + 7 + 7 + 7	6 + 6 + 6	9 + 9 + 9 + 9 + 9 + 9 + 9
2 + 2 + 2 + 2 + 2 + 2 + 2 + 2 + 2 + 2 + 2	12 + 12 + 12 + 12 + 12 + 12 + 12 + 12 + 12	5 + 5
8 + 8 + 8 + 8 + 8 + 8 + 8	5 + 5 + 5 + 5	1 + 1 + 1 + 1 + 1 + 1 + 1 + 1 + 1 + 1 + 1

REPEATED ADDITION PUZZLE
YOU NEED: GLUE SCISSORS

DIRECTIONS: Follow the directions on the next page.

8 + 8 + 8	9 + 9 + 9 + 9 + 9 + 9 + 9 + 9 + 9	1 + 1 + 1 + 1 + 1 + 1
4 + 4 + 4 + 4 + 4 + 4 + 4 + 4 + 4 + 4	7 + 7 + 7	7 + 7 + 7 + 7 + 7 + 7
8 + 8 + 8 + 8 + 8 + 8	2 + 2 + 2 + 2 + 2	5 + 5 + 5 + 5
7 + 7 + 7 + 7 + 7 + 7 + 7 + 7 + 7 + 7	10 + 10 + 10 + 10 + 10	9 + 9 + 9

REPEATED ADDITION PUZZLE

DIRECTIONS: Cut these puzzle pieces apart. Glue the piece with the matching multiplication problem on top of each repeated addition problem on the previous page.

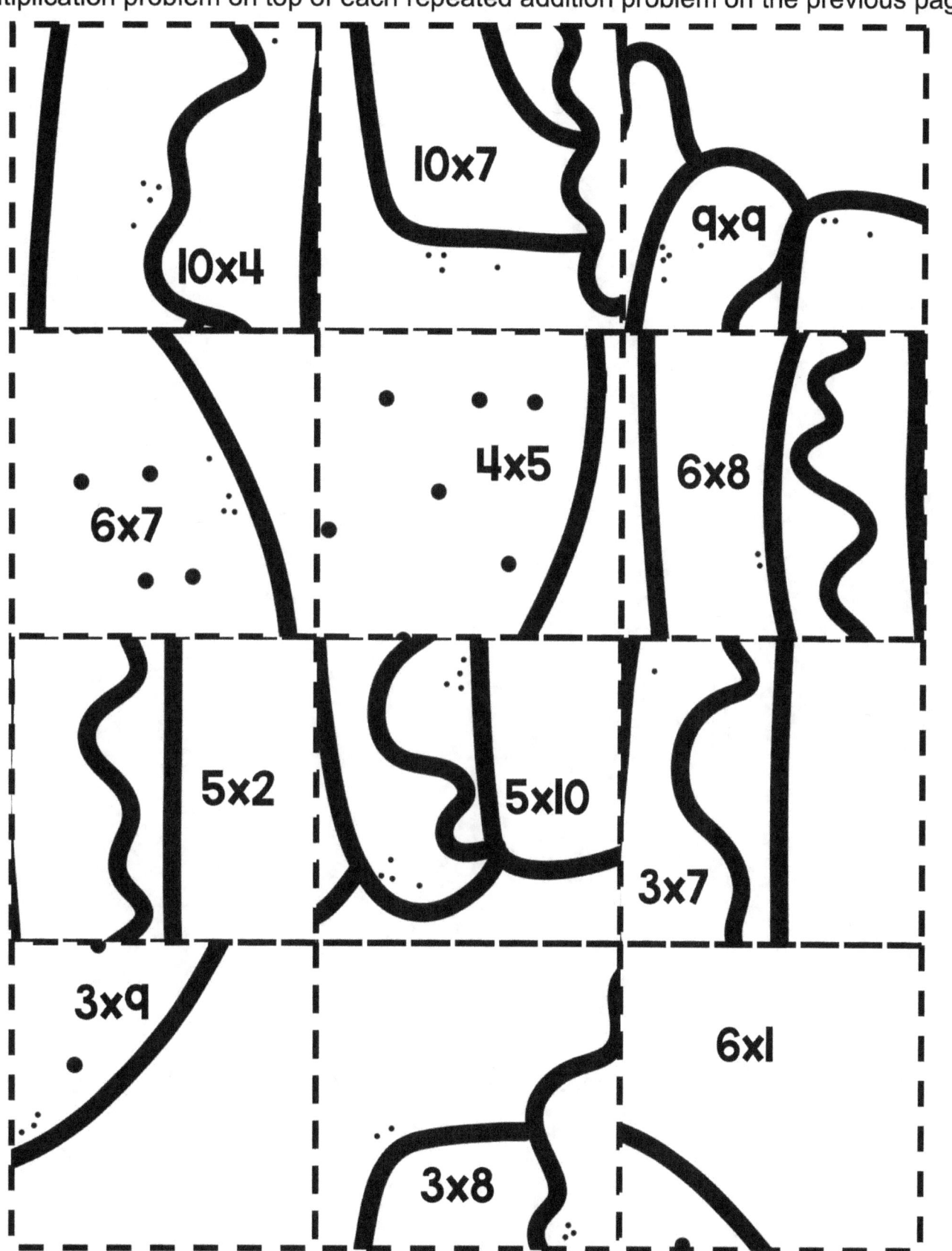

TIC-TAC-TOE: ARRAYS

DIRECTIONS: Play a game of tic-tac-toe! Before you mark a space as yours, you must name the multiplication sentence made by the array.

MULTIPLICATION ARRAYS PUZZLE
YOU NEED: GLUE 🧴 SCISSORS ✂️

DIRECTIONS: Cut the puzzle pieces from the next page apart. To solve the puzzle, glue the piece with the missing number on each set of problems. Do not cut this page apart.

©Laura Putman Digitals LLC, 3rd Grade Engaged, 2024-present All rights reserved.

MULTIPLICATION ARRAYS PUZZLE

DIRECTIONS: Cut these puzzle pieces apart. Glue the piece with the correct missing number on top of each matching set of problems on the previous page.

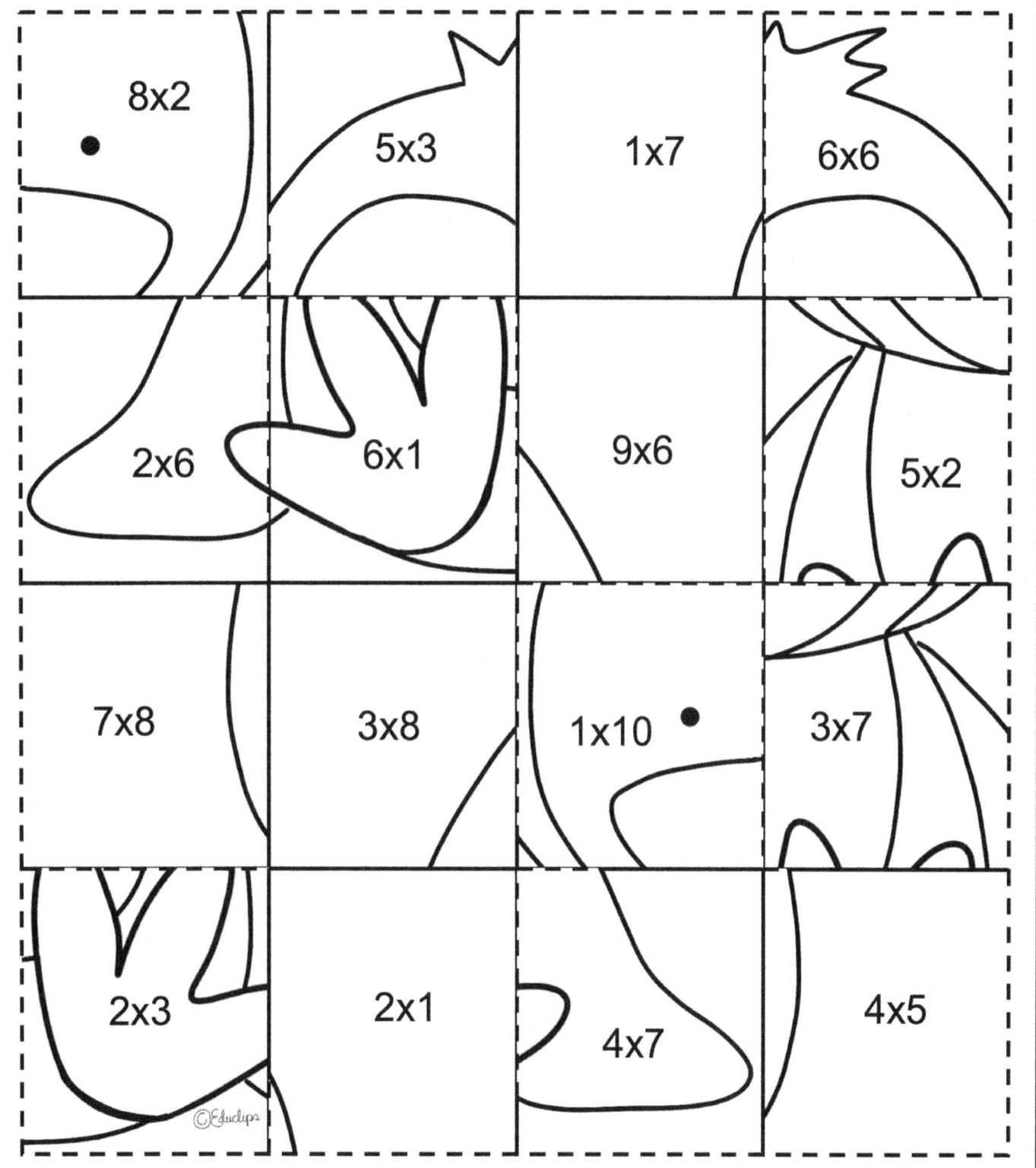

FOUR PROBLEMS IN A ROW: ARRAYS
YOU NEED: CRAYONS OR COUNTERS

DIRECTIONS: On your turn, choose an array. Say the multiplication sentence that it equals and color it or cover it. Each player uses a different color. The first player to get 4 in a row wins!

MULTIPLICATION FACTS

Parent Strategy Guide

Get the most out of this workbook!

- ✓ Tips for Success
- ✓ Strategy Posters
- ✓ Progress Chart

Scan here to get it!

The guide to helping your child master multiplication facts

By Laura Putman, M.Ed

© Laura Putman, Bright Minds Engaged, 2026-present, All rights reserved.

Does your child need extra practice in multiplication facts?

If your child is benefiting from our Long Division Workbook, you can get our **Multiplication Facts Workbook** designed to engage kids who struggle with learning multiplication facts.

Multiplication Facts Workbook

This workbook is perfect for:
- ✔ Kids who struggle with math facts
- ✔ Ages 8-12
- ✔ Reinforcing multiplication facts
- ✔ Kids who need math motivation
- ✔ Busy parents
- ✔ Math practice at home

It includes:
- ✔ A variety of pages to engage kids
- ✔ Repeated practice to build facts
- ✔ An easy-to-use answer key
- ✔ Kid-friendly graphics
- ✔ Enough practice for real mastery

Used by teachers, tutors, parents, and homeschool families to build confidence and close learning gaps.

Scan to buy on Amazon.

www.ingramcontent.com/pod-product-compliance
Lightning Source LLC
Chambersburg PA
CBHW062216220526
45471CB00009B/3220